Solar Energy for Development

Solar Energy for Development

Proceedings of the International Conference
held at Varese, Italy, March 26-29, 1979
by the Commission of the
European Communities

Martinus Nijhoff Publishers – The Hague/Boston/London   1979
for
The Commission of the European Communities

The distribution of this book is handled by the following team of publishers:

for the United States and Canada

Kluwer Boston, Inc.
160 Old Derby Street
Hingham, MA 02043
USA

for all other countries

Kluwer Academic Publishers Group
Distribution Center
P.O. Box 322
3300 AH Dordrecht
The Netherlands

ISBN-13: 978-90-247-2239-6   e-ISBN-13: 978-94-009-9332-7
DOI: 10.1007/978-94-009-9332-7

Publication arranged by
Commission of the European Communities,
Directorate-General for Scientific and Technical Information and Information Management,
Luxembourg

EUR 6377 EN, FR

For further information:
Martinus Nijhoff Publishers b.v., P.O.B. 566, 2501 CN
The Hague, The Netherlands

LEGAL NOTICE

Neither the Commission of the European Communities or any person acting on behalf of the Commission is responsible for the use which might be made of the following information.

Preface

The International Conference "Solar Energy for Development" was held from the 26th to the 29th of March 1979 in Varese, Italy. The Conference was organised by the Commission of the European Communities to assess the potential of solar energy for meeting the needs in the developing countries, particularly in their rural areas. The objectives of the Conference were threefold:

— To review those solar energy technologies which are appropriate for large scale utilisation in the short and medium term;

— To appraise problems which may be alleviated by a better use of conventional solar energy and the introduction of appropriate new solar technologies;

— To recommend ways and means of extending the use of solar energy, taking into account technical and non-technical criteria.

Before the Conference, in September and October 1978, five regional solar energy seminars were held in Nairobi (East Africa), Bamako (West Africa), Amman (Arab countries), Caracas (Latin America) and New Delhi (South and South-East Asia). With the help of the experts at these seminars a general working document was established and made available to the participants of the Conference. 280 experts from 80 countries all over the world were invited by the Commission to attend the Conference. The United Nations and 11 other regional and international organisations were represented.

The Conference was opened by Dr. Guido Brunner, Member of the Commission responsible for Energy, Research, Science and Education, in the presence of Dr. Antoniozzi, Minister for the coordination of scientific and technological research of Italy, and other high officials. The current development policy of the Commission of the European Communities was presented by Mr. Maurice Foley, Deputy Director General for Development.

The Conference was organised as a forum for exchange of information and constructive dialogue. Each session was opened by a keynote speech followed by free discussion. This report includes the full text of all keynote speeches and summaries of the discussions prepared by rapporteurs.

The executive summary which precedes the mainbody of this report is an amended version of a draft submitted to the Conference in its final session which takes into account the suggestions expressed on that occasion.

This report will be submitted to the UN Conference on Science and Technology for Development which is to be held in August 1979 in Vienna. No doubt, it will also be of use to the Commission of the European Communities and the participating countries and organisations as a guide to their future programmes and actions on solar energy.

The organisers take this opportunity to thank again the chairmen, rapporteurs, keynote speakers and all experts whose active and constructive contribution made possible the success of the Conference.

G. SCHUSTER
Director General for Research,
Science and Education
Commission of the European Communities

Table of contents

I. Executive summary

CONTENTS

1. INTRODUCTION: RATIONALE AND OBJECTIVES

Solar energy holds out a large potential for *all* countries. It appears to be among the most promising energy options for developing countries, especially in the rural .areas. This is due to:

— particularly favourable conditions in terms of solar energy input (direct radiation, biomass, wind, etc.);
— decentralized needs not involving the infrastructures of conventional energy networks;
— consistency with some existing technological capabilities: several solar energy techniques such as passive heating and cooling or biomass have been extensively used for ages. They require only improvements.

In the medium and long term, the main impact of a large utilization of solar energy in the developing countries will be the achievement of the following objectives:

— to facilitate further development of the countries in the field of living conditions, food production, development of labour skills and employment;
— to reduce their dependence on imported energy and enable them to attain the maximum possible degree of energy self-sufficiency.

2. GUIDELINES

There is general agreement that regional and international cooperation is highly desirable to develop solar energy. Hence, developing countries should cooperate on an equal basis among themselves and with the industrialized countries. Cooperation should include joint design of projects, complementary production in the developing and industrialized countries by using existing capabilities, and the development of indigenous skills.

Developing countries should not become just consumers of solar energy technology but should be involved at all stages of its development and use. Development of solar energy also on a national level is desirable. This development should be part of a national plan to develop the indigenous energy sources. Solar energy may help to create a technological attitude in rural areas. Development should in all cases be based on a prior analysis of the site-specific needs and conditions of use which exist in the various regions and in particular countries.

3. TECHNICAL ISSUES

3.1. Radiation and wind data

In many countries there is equipment for radiation and wind measurements, but, frequently this equipment does not cover all important areas adequately.

For monitoring purposes, the existing meteorological equipment is sometimes too expensive and co-ordinated efforts should be made to improve the present situation.

Therefore, priority should be given to the following:

— development, in collaboration with WMO, of regional stations, well equipped and serving as local agent for collecting the data and distributing this information as well as relevant guidance material;

— correlation of the radiation data of the regional stations with satellite data in order to get more reliable information on the spatial distribution of solar radiation;

— development and testing of cheap devices for the monitoring of solar radiation and wind implying simple maintenance, data acquisition and data handling.

3.2. Biomass

Biomass represents a renewable and large source of energy. Many developing countries have an abundant supply of it, with climatic conditions favourable to its exploitation.

There are many ways in which it can be used to produce energy and several countries have already exploited them to a significant extent for instance, for industrial production of alcohol from sugar cane by fermentation-distillation, which offers large possibilities for the substitution of fossil fuels.

In Asia, there are millions of digesters for biogas production with many uses such as domestic cooking and heating, lighting, fueling of engines, agro-industry. Electricity production through generators employing biomass also offers great possibilities in the rural sector. It is important to note, however, that biomass fuels should always be used in such a way as not to interfere with food production; its consumption rate should not exceed its rate of growth.

Owing to its possible large impact in the short term, a first-ranking priority must be given to the development and optimisation of processes related to its exploitation and to research aimed at accelerating growth of biomass.

3.3. Water

Water pumping

Experience has already been obtained under field conditions. Both photovoltaic and thermodynamic systems have given in most cases encouraging results although costs are at present high. There is a potential for extensive use of solar pumps in rural areas if cost reduction and further improvements in performance can be achieved. Because of the importance of irrigation in agriculture, high priority should be given to this application as well as for cattle breeding. Urgent attention should also be given to water needs in the villages. Wind powered pumps are still extensively used in some regions. They are attractive for areas with strong and regular winds, if regular maintenance can be provided.

Desalination

Solar desalination using simple technologies has been employed in many regions, although difficulties have been encountered, mainly associated with corrosion and maintenance. Because of the important need for fresh water in some areas, further research and development is advisable.

Water conservation (greenhouses)

The use of plastic greenhouses in arid areas has been successful in increasing productivity while greatly reducing water consumption. Present success should continue as it is believed that widespread application will be economic.

3.4. Power production

1. Solar thermal power conversion in the range above 1 kW has been used in a number of plants. They are promising for further applications provided reliability in operation and maintenance can be improved. They may be operated in association with thermal storage systems. Thermal solar power generators and the associated collectors and subsystems should be produced to a large extent by local manufacturers to reduce costs. Experience with turbines and piston engines should be increased to elaborate optimum solutions.

2. Photovoltaic power generators have a large potential for cost reduction. They can be used at low power levels (1 W - 1 kW) for lighting, communication systems, battery chargers but also offer promise in larger sizes. Maintenance-free operation is one of the main advantages. More field testing of photovoltaic systems is required to demonstrate its potential under different climatic conditions. Priority should be given in R & D to reduction of production cost of cells.

3. Biomass can be employed for power generation by means of combustion engines. In this way it is attractive for both stationary power and transportation. Besides conventional or modified heat engines, the highly developed and cheap car engine is suitable. Agricultural wastes and residues are attractive raw materials for fuel production and offer economic solutions in the short term.

4. Windmills and small-scale hydro power generators are considered especially useful for rural areas. They are based on traditional use and experience, imply locally available materials and manufacture and thus offer the most favorable price/power ratio.

5. Storage of electrical power deserves due attention in particular in association with photovoltaic and wind power generators. Available storage batteries need improvement in terms of capacity, service and cost.

As solar radiation and wind availability are often complementary, the association of a wind generator with a photovoltaic generator is attractive to reduce the need for storage and to reduce cost provided the local radiation and wind conditions are favourable.

6. All types of power generators demand extensive field testing under local conditions. Considerable reduction of cost of all existing solar power systems has to be achieved.

3.5. Heat

Cooking

Cooking in rural areas represents the largest energy need and deforestation is becoming critical in some areas. This leads in many cases to progressive desertification. Highest priority should be given to this problem. The widespread introduction of improved conventional cookers, for which the technology already exists, would be a low-cost and immediate contribution to its solution. Reforestation with fast growing species together with planned exploitation of existing forests should also be treated with high priority.

Solar cookers employing direct solar radiation have been developed but their introduction has not been successful on a significant scale.

Biogas cookers have been demonstrated successfully and offer considerable potential. Biogas production still leaves considerable room for technical development.

Refrigeration

The preservation of food and medicine by refrigeration could be of benefit. Solar cooling techniques are under development, but considerable cost reductions are a prerequisite for their application.

Heating and cooling

Optimum design of building, using in many cases traditional architecture, is seen as the best approach to reducing energy demand. Solar water heating systems are already manufactured in many of the developing countries and their wider use may be beneficial.

Solar energy in Agriculture and Industry

Solar energy can contribute in many ways to energy self-sufficiency of agriculture and agro-industry. In particular solar crop drying has the potential for great benefit. Solar process water heating may also be employed.

Plastic greenhouses are extremely successful and are of paramount importance for the future. In hot countries solar heat may be used for soil decontamination. Solar heat has also potential as process heat in the agro-industry, cottage industry and other industry as well.

Solar power deserves also consideration in agriculture and industry e.g. for grain grinding in the villages.

Demonstration projects at village level are highly desirable.

11

4. GENERAL IMPLICATIONS

A number of considerations arise from the implementation of the technical solutions discussed in section 3 and from the experience of participants. Some of the more important are listed below.

4.1. Development of technological capabilities

It is desirable that, as far as possible, solar energy equipment should be designed and manufactured locally since, in this way in addition to providing employment, such equipment can be more easily erected, regularly maintained and repaired. Such industry should make use of local skills and materials and also locally available finance (as far as possible).

The industry should be encouraged and assisted with the framework of a national programme to provide market stability and continuity. An institutional back-up for research, development and quality control are also required and standardisation of equipment and specification is necessary.

Technical co-operation is essential to improve the autonomy of the developing countries. This can be achieved, provided that such a local infrastructure for development and production is first instituted. Technology should be encouraged to grow indigeneously.

4.2. Economic aspects

With respect to economics, several specific requirements arise:
— local manufacture should be used wherever possible to improve the competitiveness of solar technology;
— in comparing solar equipment with other devices, account should be taken of the social costs and benefits for the community;
— the enduser should make some personal investment in the equipment.

4.3. Information

A great deal of scientific and technical information exists and there is a need to develop methods for its wide dissemination. Far less is known on the socio-economic benefits of actual field use. Clearly, more information is necessary on the needs and preferences of the potential users.

International channels are required for the transfer of information in both directions: intensive national information and demonstration programmes should assure the transfer of existing know-how and information on new developments from research institutes to village level.

The awareness of local authorities and possible users of the potential of solar energy may be increased by a proper information policy.

4.4. Education and training

There is a requirement for training at all levels, but particularly at technician level to provide personnel for design, manufacture, installation and maintenance. This implies the provision of local training capabilities. Specialist courses at University level, in-field training and the introduction of suitable material in secondary and primary school curricula should all be considered.

Regular information exchange on practical experience should be arranged in regional seminars.

4.5. Social and environmental conditions

Solar technologies will be useful only if they are socially and culturally acceptable. At the same time such technologies have to be developed in such a way that they do not result in environmental problems. Considerable field work is required to establish information on social acceptability.

4.6. Research and development

Many solar technologies are still far from being cost competitive. As it is clear, however, that at least some of them will eventually be very important for the future it is absolutely essential to support and strengthen their development. New ideas and applied and pure research on solar energy deserve encouragement.

5. RECOMMENDATIONS

1. All countries, developed and developing, should be encouraged to undertake comprehensive energy surveys and prepare short and long term energy development plans, in which the proved and potential uses of renewable sources of energy, in particular solar energy, are given full consideration. These plans should be co-ordinated with and integrated into rural development programmes and should take account of real needs and the social, environmental and cultural acceptability of the proposed solutions. Assistance in the preparation of energy plans should be made available to developing countries by the developed countries, inter-governmental organisations and international agencies, preferably but not exclusively under existing collaborative arrangements.

2. The developing countries, with the collaboration and help of developed countries, intergovernmental organisations and international agencies, should be encouraged to set up and strengthen the indigenous infrastructure needed to undertake or extend their energy plans. Education and training at all levels is an essential element.

3. More effort should be allocated to in-field testing and demonstration projects, embracing all stages from R & D to final installation, monitoring and technical follow-up. The developing country concerned should be involved in all stages

from the initial proposal onwards. Specific programmes for experimental life testing of components should be set up. The controversial subject of integrated projects should be carefully studied to determine its real potential.

4. Existing international channels should be used to exchange and disseminate multilingual information on the development of solar energy technology, infield testing, demonstration projects (including negative results) and the progress of implementation. The information services should include periodic newsletters and instructional hand-books. It must be stressed that the writing style of this material should be simple and easy to understand.

5. Existing formal links between national institutions with an interest in solar energy should be strengthened and new links established to promote collaboration in this field of technology.

6. The Commission should set up advisory bodies to strengthen co-operation between the developed and developing countries.

7. Arrangements should be made for the exchange of workers in R & D, education, vocational training, manufacture (including local manufacture), installation, testing, monitoring and maintenance between developed and developing countries and between developing countries themselves. In this connection, the training and advisory roles of the Joint Research Centre, Ispra in the solar energy field should be expanded.

8. The Commission should periodically organise regional workshops to exchange information on progress in specific areas of solar energy technology.

9. The Commission, in collaboration with the developing countries, should organise a second conference on Solar Energy for Development in about two years' time to assess progress.

II. Conference Agenda

MONDAY, 26 MARCH 1979

Opening Session,

Chairman
Dr. G. SCHUSTER, Commission of the European Communities, Director General for Research, Science and Education

Opening Address
Dr. G. BRUNNER, Member of the Commission of the European Communities, Responsible for Energy, Research, Science and Education

Address of Welcome
Dr ANTONIOZZI, Minister for Coordination of Scientific and Technological Research, Italy

Address of Welcome
Dr. Proc. G. GIBILISCO, Mayor of Varese

Keynote Speech: Solar Energy for Development
Sir George PORTER, F.R.S. Director of the Royal Institution of Great Britain

Film: The Energy Problem of an Indian Village
Dr. H.N. SHARAN, Director, Bharat Heavy Electricals Ltd, India

Presentation of a Summary of the Preparatory Seminars
Dr. W. PALZ, Commission, Directorate General for Research, Science and Education

Session 1: Water,

14:30 - 18:00

Prof. S. QASEM, Dean, Faculty of Agriculture, Jordan

Co-chairman
Dr. A. STRUB, Commission, Directorate General for Research, Science and Education

General Rapporteur
Mr. P. LEQUEUX, Commission, Directorate General for Development

(1.1) Keynote Speech
Dipl. Ing. F.J. FRIEDRICH, Project Management for Non-nuclear Energy Research, Kernforschungsanlage Jülich, Federal Republic of Germany

| (1.2) | Introduction to the Working Document on Water |
| | Mr. P. LEQUEUX, Commission, Directorate General for Development |

(1.2) — Introduction to the Working Document on Water
Mr. P. LEQUEUX, Commission, Directorate General for Development

(1.3) — Specialized Technical Review Meetings

(1.3.A) — **Water Pumping,**

Chairman
Prof. R. VAN OVERSTRAETEN, Katholieke Universiteit Leuven, Belgium

Rapporteur
Prof. P. DUNN, International Technology Development Group, United Kingdom

(1.3.B) — *Water Desalination,*

Chairman
Dr. F.A. DAGHESTANI, Deputy Director General, Royal Scientific Society, Jordan

Rapporteur
Mr. B. DEVIN, Commissariat à l'Énergie Atomique, France

(1.3.C) — **Greenhouses to Save Water,**

Chairman
Dr. P. CHARTIER, Director, National Institute for Agronomical Research (INRA), France

Rapporteur
Dr. A. STEHLER, Technical University Munich-Weihenstephan, Federal Republic of Germany

(1.4) — General Discussion and Recommendations for Future Actions

TUESDAY, 27 MARCH 1979

Session 2: Power Production,

9:00 - 12:30

Chairman
Dr. C.L. GUPTA, Director, Tata Energy Research Institute, India

Co-chairman
Mr. D. VINCENT, Commission, Directorate General for Development

General Rapporteur
Prof. W.H. BLOSS, University of Stuttgart, Federal Republic of Germany

(2.1)   Keynote Speech
Dr. H. DURAND, President, Commissariat à l'Énergie Solaire, France

(2.2)   Introduction to the Working Document on Power Production
Prof. W.H. BLOSS, University of Stuttgart, Federal Republic of Germany

(2.3)   Specialized Technical Review Meetings

(2.3.A)   **Thermal Power Generation,**

Chairman
Dr. A. STRUB, Commission of the European Communities

Rapporteur
Dr. J. GRETZ, Joint Research Centre of the European Communities, Ispra

(2.3.B)   **Photovoltaic Power Generation,**

Chairman
Dr. T.K. BATTACHARYA, Central Electronics Ltd, India

Rapporteur
Dr. K.H. KREBS, Joint Research Centre of the European Communities, Ispra

(2.3.C)   **Biomass Fuels for Stationary Power and Vehicles**

Chairman
Dr. S.C. TRINDADE, Director, Centro de Tecnólogia Promon, Brasil

Rapporteur
Mr. J. PHELINE, Commissariat à l'Énergie Atomique, France

(2.3.D)   *Wind and Small Scale Hydraulic Power*

Chairman
Dr. P. MUSGROVE, University of Reading, United Kingdom

Rapporteur
Dr. F. JÄGER, Institut für Systemtechnik und Innovationsforschung, Federal Republic of Germany

(2.4)   General Discussion and Recommendation for Future Actions

Session 3: Solar Heat,

14:30 - 18:00

Chairman
Dr. P. della PORTA, Vice President, SAES-Getters, Italy

Co-chairman
Prof. Z. TABOR, Israël

General Rapporteur
Mr. B. MEUNIER, Sema, France

(3.1)   Keynote Speech
Prof. A.A.R. ELAGIB, Chairman, Scientific and Technological Research Council, Sudan Engineering Council, Sudan

(3.2)   Introduction to the Working Document on Solar Heat
Mr. B. MEUNIER, Sema, France

(3.3)   Specialized Technical Review Meetings

(3.3.A)   *Cooking,*

Chairman
Dr. R.K. SURI, Bharat Heavy Electricals Ltd, India

Rapporteur
Mr. T.C. STEEMERS, Commission, Directorate General for Research, Science and Education

(3.3.B)   *Cooling Chambers, Ice Production, Refrigeration*

Chairman
Dr. H. TREIBER, MBB, Federal Republic of Germany

Rapporteur
Dr. B. McNELIS, General Technology Systems Ltd, United Kingdom

(3.3.C)   *Heating and Cooling in Habitat,*

Chairman
Prof. A. MOUMOUNI, Director, Office de l'Énergie Solaire, Niger

Rapporteur
Dr. E. ARANOVITCH, Joint Research Centre of the European Communities, Ispra

| (3.3.D) | *Solar Energy in Agriculture,* |

(3.3.D)   *Solar Energy in Agriculture,*

Chairman
Prof. D. HALL, King's College, United Kingdom.

Rapporteur
Dr. L. CUOZZO, Director, CTIP Solar, Italy

(3.4)   General Discussion and Recommendation for Future Actions

WEDNESDAY, 28 MARCH 1979

Session 4: International and Regional Cooperation,

9:00 - 12:30

Chairman
Dr. A. STENMANS, Secretary General, Programmation de la Politique Scientifique, Belgian Government

General Rapporteur
Dr. R. BATTI, Commission, Directorate General for Research, Science and Education

(4.1)   Keynote Speech
Dr. F.A. DAGHESTANI, Deputy Director General, Royal Scientific Society, Jordan

(4.2)   Introduction of the Working Document on International and Regional Cooperation
Dr. R. BATTI, Commission, Directorate General for Research, Science and Education

(4.3)   Specialized Technical Review Meetings

(4.3.A)   *Radiation Data,*

Chairman
Prof. J.K. PAGE, University of Sheffield, United Kingdom

Rapporteur
Dr. J.W. GRÜTER, Kernforschungsanlage Jülich, Federal Republic of Germany

(4.3.B)   *Research, Exchange of Information,*

Chairman
Dr. M. CRESPI, National Atomic Energy Commission, Argentina

Rapporteur
Prof. F. FITTIPALDI, University of Napoli, Italy

(4.3.C)   *Industrialisation, Transfer of Technology,*

Chairman
Dr. ARMANI, Centre for Industrial Development, Brussels

Rapporteur
Dr. A. REITHINGER, Commission, Directorate General for Development

(4.3.D)   *Education, Training,*

Chairman
Prof. P.M. GITHINJI, University of Nairobi, Kenya

Rapporteur
Dr. G. LIVI, Commission, Directorate General for Development

(4.4)   General Discussion and Recommendations for Future Actions

Session 5: Environmental and Social Implications

14:30 - 18:00

Chairman
Prof. P. DE MEESTER, Katholieke Universiteit Leuven, Belgium

Co-chairman
Dr. J.C. RENAUD, Chief Adviser, Commission, Directorate General for Energy

General Rapporteur
Dr. G. BEGHI, Joint Research Centre of the European Communities, Ispra

(5.1)   Keynote Speech: Environment and Social Implications
Dr. D. FALL, Director of IPM, University of Dakar, Senegal

(5.2)   Introduction of the Working Document on Environmental and Social Implications
Dr. G. BEGHI, Joint Research Centre of the European Communities, Ispra

THURSDAY, 29 MARCH 1979

Closing session

14:30 - 17:00

> Chairman
> Dr. G. SCHUSTER, Commission, Director General for Research, Science and Education
>
> Co-chairman
> Prof. S. VILLANI, Commission, Director General of the Joint Research Centre of the European Communities
>
> Rapporteur
> Dr. W. PALZ, Commission, Directorate General for Research, Science and Education
>
> Presentation of Conclusions of the Conference
>
> General Debate
>
> Closing Address

CONFERENCE ORGANISATION

Organiser

Commission of the European Communities

Conference Committee

Co-Chairmen: Dr. G. SCHUSTER,
Director General for Research, Science and Education.
Dr. K. MEYER,
Director General for Development.

Directorates General represented:

— DG I   External Relations
— DG III   Internal Market and Industrial Affairs
— DG VIII Development
— DG XII Research, Science and Education
— DG XIII Scientific and Technical Information and Information Management
— DG XVII Energy
— Joint Research Centre

In cooperation with

— ACP Group of States
— Centre for Industrial Development

Organising Group

Dr. W. PALZ, General Organiser
Mr. P. LEQUEUX, Organiser for ACP countries
Dr. G. VALENTINI
Mr. W. MARTIN
Dr. R. BATTI
Mr. F. TREBLE
Mr. R. GICQUEL

Local Organiser

Mrs. M.P. MORETTI
Centro Comune di Ricerca
I-21002 Ispra (Varese)

III. Opening Addresses, Addresses of Welcome

OPENING SPEECH

by

Mr. Guido BRUNNER,
Member of the Commission of the European Communities

Minister, Mr. Mayor, Chairman, Ladies and Gentlemen,

I am very pleased to be in Varese and to welcome you to this conference on solar energy for development. In particular, I extend a special welcome to those who have come here from outside Europe.

Italy is the obvious location in the European Community for this sort of gathering. The warmth not only of its sunshine but also of its people is legendary. Both have been rightly praised by poets and visitors over the ages. I am sure that now a group of solar energy experts will be adding their voice to the chorus.

Since we have the pleasure of being in Italy, it is an extra pleasure and an honour that we have amongst us Dr. Antoniozzi, the Italian Minister for the Co-ordination of Scientific and Technological Research.

I should also like, in opening the proceedings, to welcome the Mayor of Varese, Dr. Gibilisco. The city authorities and the chamber of industry and commerce have been of invaluable help in organizing this conference. You can imagine the problems involved in arranging an event of this size.

This is the first time that the European Commission has organized a worldwide conference specifically geared to the energy needs of developing countries. I hope this initiative constitutes a concrete step towards closer cooperation between the many and varied countries and organizations which you represent here.

It may be helpful if very briefly I try to put the role of solar energy into a broader context. The 1973/74 oil crisis demonstrated all too clearly how dependent we all are on stable and adequate energy supplies at prices we can afford. People now realize how energy underpins all energy or economic and social activities. The oil price rises of 1973/74 caused a shock to the world economic and monetary system from which we still have not recovered. But those events also made everyone think about new energy supply strategies, new production techniques, and new approaches to energy use.

Recent events in the Iran have again demonstrated the delicate balance of the present energy supply system. However, the word "energy crisis" has to be used with great care. The real problems are to be found in the underlying long term supply and demand position. It was clear before Iran production declined that in the mid- or late 1980's incremental world oil demand could outstrip incremental OPEC production capacity. That was and remains the essence of the energy crisis. What the Iranian situation has done is to remove a degree of flexibility in our supply strategies. Particularly if other Middle Eastern producers revert to lower productions levels, we shall have less time to achieve an orderly transition from oil to other energy sources. Hence the emphasis in the Community on domestic production of fossil fuels, on nuclear power, on developing new sources, or on opening up alternative sources of supply throughout the world. Hence the direct relevance of the European Council's decisions to limit oil consumption to 500 m. tonnes in 1979, and to limit oil imports to 470 m. tonnes in 1985.

The global implications of all this are serious. World demand for energy is going to increase enormously in absolute terms for the rest of this century and beyond. This is true not only of the industrialized countries but also of the developing countries. They will require a bigger percentage increase if their urgent economic goals are to be met. Even then, their per capita consumption will be about one-tenth that of the industrialized countries.

In the face of the limits on future world oil supply, what is to be done? Are we to see a competitive scramble for limited supplies? We should all be losers in that—but particularly the developing countries. The weakest would go to the wall.

It is our duty to find a better way. We must recognize the increasing interdependence which sustains all countries, east or west, north or south, rich or poor. We are all in the same boat. Realizing that we all have a common interest in economic survival, we must work out a responsible and orderly management of scarce world resources.

Therefore energy consumption must be curbed—but selectively, so that the chances of world economic recovery are not destroyed. And the gap between rich and poor countries must be reduced—not by levelling down, but by levelling up.

This is very good in theory. To achieve it in practice requires concrete action. In industrialized countries there is an urgent need for energy saving, and a radical revision of our attitude towards energy use, both short term and long term. Also there are the financial resources and technical expertise to switch into non-oil resources. Massive efforts on coal and nuclear are a priority—but oil will remain a dominant fuel. In the longer term, we must look towards new sources—including solar energy. In spite of concentrated efforts on new sources in the European Community, they may not supply much over 5% of our total supplies in 2000. This is of course, a large amount of energy—but the small share indicates the scale of the problem.

In the developing countries, the perspectives are different. There, totally new economic structures are being created. It is essential to build in to these new investments, the most modern energy saving techniques. Demand is often less centralized or concentrated, more scattered on a smaller scale. This demand pattern is inherently better suited to the use of new sources. In addition, the climate is such that solar energy, for example, has exciting new possibilities. It is outside the large cities that solar power seems to offer the most promising solution to some of the Third World's energy problems. The problem of rural development, the equipping of villages that are isolated and with poor communications has inspired much research into a source of energy suited to such local conditions.

At the same time, the conventional energy resources of the developing countries are probably vast, but still largely uncharted. The Community could help to survey these resources and put them at the service of the countries concerned. In this way, the energy situation of the world should be improved.

There is therefore a clear and important link between the world energy situation, the position of the industrialized countries, the position and needs of the developing countries, and the long term role of new energy sources.

The Community's policy towards the developing countries is well known. We have the so called Lomé agreement, an extensive general agreement for cooperation with 56 developing countries known as the ACP—that is African, Caribbean and Pacific countries. It covers all economic sectors and serves also as a framework for cooperation on solar energy with these countries. Solar energy will also pay an increasing role in the Community's agreements for cooperation with other groups of States, such as the Mediterranean countries, ASEAN, and the countries of Latin America.

Outside these specific agreements, the Commission wants to make a bigger effort in this field. We want to establish a new programme of energy cooperation with the non-oil producing developing countries. We see scope in bilateral action, action through multilateral agencies, and action with the oil producers. OPEC itself has an impressive record of overseas aid. We also hope to be able to contribute to the efforts of regional energy research centres in developing countries — such as OLADE in South America.

The first step would be to draw up with selected developing countries inventories of their energy requirements and resources. This would be backed up by aid for prospecting, and later production technology, training etc. This would help the developing countries solve their own energy problems, and ease the pressure on world supplies.

Science and technology generally have assumed a high importance in the process of development. I am happy to say the European Community has involved itself in the many discussions presently taking place on this subject, and in particular in relation with the United Nations Conference on Science and Technology for Development.

I should now like to say a few words about the Community's involvement in solar energy research. Appropriately enough, it is in Italy that we have initiated our largest single project in solar energy research. A demonstration solar tower, of one megawatt capacity, is being constructed in Sicily by an international consortium of industries. We are providing half of the finance for this. It will supply electricity into the Italian grid in about 2 years from now.

More broadly, we are trying to set up a plan of action aiming at the better use of solar energy not only in the Community, but also in the developing countries and in particular in their rural areas. We hope to make a comprehensive review of solar energy technologies which are appropriate for wide scale use in the short- and the medium-term. We aim to identify the problem and situations which may be improved by a better use of conventional solar energy and by the introduction of appropriate new technologies.

The European Communities have been active in the field of solar energy research for a long time. We carry out inhouse research at our Joint Research Centre at Ispra. It is verly close to here and I understand you will have a chance to visit it later in the week. We also sponsor and fund some 250 research projects in all our Member Countries. The solar tower in Sicily which I mentioned earlier, is only one example. There is also considerable effort underway in fields like photovoltaic power generation, solar heating and cooling, biomass and others. We are now also stepping into the field of solar energy demonstration projects. This is part of a broader programme of almost $ 200 million over four or five years which we plan to devote to demonstration for energy saving processes and for the exploitation of new sources including solar, geothermal and liquefied and gasified coal.

At this Conference we want to favour exchange of information so that everybody can benefit from existing experience. Much emphasis has therefore been put on discussion time, rather than on formal presentation of technical papers. As you represent such a diversified spectrum of technical competence, and as your countries show specific needs that cover a wide range of problems, your experience and your assessment in this field are likely to be very different from each other. I intend to submit the conclusions of this Conference to UNCSTD in Vienna next August, and I hope to be your spokesman when stressing that solar energy should become a priority area in any resolutions or recommendations that may ensue.

Ladies and Gentlemen, I am sure that your deliberations will make a substantial contribution to the future progress of solar energy use. I sincerely hope that you will go home with an enhanced knowledge and experience, and that this will help you in your efforts to utilize the most abundant and renewable source of energy for the benefit of mankind.

Ladies and Gentlemen, it gives me great pleasure to declare open the Conference "Solar Energy for Development". I wish you all an enjoyable and efficient meeting.

ADDRESS OF WELCOME

by

Mr. ANTONIOZZI,
Minister for the Coordination of Scientific and Technological Research, Italy

Mr. Chairman, Mr. Brunner, Ladies and Gentlemen,

It is a very great pleasure for me to present the greetings of the Government of Italy to this conference and to extend my gratitude to all those who accepted the invitation of the Commission of the European Communities to participate.

In my capacity as Minister for Scientific Research, I attach the greatest importance to the work of this Conference, which has been so competently initiated by Commissioner Brunner, who is responsible for energy, research, science and education and whose merits of having promoted these fields are now established.

I wish to address Mr. Brunner as well as the two co-chairmen, Messrs. Schuster and Meyer, to convey to them the satisfaction of the Government of Italy as well as my own; and I thank them particularly for having selected this town and this region which, thanks to their beautiful and historical sites, contribute to creating a spirit that will favour the exchange of opinions and experience on subjects of particular topicality and of special interest to the progress of civilization.

The topics you will be discussing will contribute not only to the autonomy of each country as far as the organization of its progress is concerned but also to the changes to be adopted for the current development schemes of our countries and that of the European Community. The discussions will also reveal the changes necessary to correlate the Community's development with that of other peoples and regions of the world.

This conference thus also represents an opportunity for us to become aware of the needs of other peoples and to understand that progress does not mean imitating other models of life. Cooperation stands for applying our techniques to the local circumstances and changing the prevailing conditions of life, that hamper the development of society.

It is within this sphere that we have to define the support we are able to give to the development of other peoples and it is within this framework that Italy feels the cooperation with developing countries to be a **full component** of its international economic relations and in accordance with its principle of the interdependence of the recent development of all countries, as is being confirmed by the law passed on February 9, 1979.

By virtue of this law, Italy is committing itself, in agreement with the European Economic Community and other international bodies, to promoting economic and social, technical and cultural progress of all countries, with respect to their individual development programmes and in a spirit of solidarity with peoples as well as in accordance with the principles established by the United Nations.

The activities in the fields of cooperation with developing countries among others will be:

— defining and carrying out development projects, in particular in the field of agriculture, energy, industry and crafts, infrastructures, health services, social and culture, tourism, scientific and technical research;

— promoting and ranting credits and other financial support to central banks and public bodies in developing countries, to support the aims and cooperative programmes in which Italy is participating and which these countries approved and actually promoted;

— participating, also financially, in the activities of the Community and international bodies and funding cooperation with developing countries;

— assisting the peoples of the lesser developed countries afflicted by natural calamities and facing particularly strong and urgent need;

— promoting cultural exchanges between Italy and developing countries especially as far as young people are concerned;

— multiplying the instruments of, and promoting the initiatives undertaken for the qualification and the detachment of young people as co-operators to developing countries.

These forms of support within the aforementioned law include, as a matter of fact, scientific research and it may therefore be considered to be one of the most modern laws on technical cooperation.

Since the present Conference is on solar energy, will you allow me to recall that Italy decided some time ago to intensify its own commitment in research in this field?

The importance of developing solar energy has been established by the National Council on Scientific Research, which provided financial support amounting to 3 billion 250 million lire for the period 1976 to 1978. About 4 billion lire will be invested by the Council in this field in the course of 1979. Additional investment is envisaged in industrial research in this sector from ad-hoc funds for applied research.

The great number of uses of this new energy will be one subject of your discussion and exchanges of experience. As far as I am concerned, I am content to stress that energy is an essential element in the development of every country but environmental conditions in some countries are such as to favour the application of new forms of energy. This is particularly true of agrarian societies in certain regions of the world.

In some areas it might be feasible to install facilities using the sun as energy source, for example where little energy might be sufficient to put in operation irrigation systems in order to create artificial farming-soil thus altering the nature of the ground.

Solar energy can be used not only for illumination but also for heating, drying and distillation purposes. These uses are of great interest to certain agrarian economies and are capable of changing the living conditions of man.

The subject of this Conference appears to me particularly interesting with reference to present problems and I am confident that the information you will exchange, as well as the experience you will gain and have the opportunity to compare, will represent an essential contribution to ideas that might help the process of political decision-making in the participating countries.

Thank you.

ADDRESS OF WELCOME

by

Dr. Proc. G. Gibilisco,
Mayor of Varese

Ladies and Gentlemen,

It is with great pleasure that I am giving you, as Mayor of Varese, the welcome of this town. But, before a welcome, let me thank you for choosing our town for such an important and considerable congress. Many thanks for your presence but above all many thanks for the subject you chose to develop during these days.

We are living in a society utilizing energy more and more. Beyond the international political problems, this society feels today an urgency, that cannot be put off, to go on through new ways. Should these new ways not be followed, not only the possibility of further development but also the existence of all the achievements of men, would be endangered.

The subject you chose is therefore a cultural problem before being a technical and scientific problem, and this problem puts serious questions to the man about the future that is going to be planned for him. It is therefore necessary that the world, thanks to the peaceful coexistence of the countries, be united and pursue a future for the man, who is the same in any latitude. It is necessary to blame the waste of energy devoted to the war, because through the war the scientific progress leaves behind itself only fear in the heart of men. Doesn't the fear of many people of nuclear energy originate from war? But nuclear energy is for peace.

Today, the European Communities, which is acquiring more and more importance, particularly during this year through the European elections, is an instrument of peace, giving therefore to the peoples the possibility of promoting the progress and the future of men.

Let me now, speak about our dear town of Varese, about its geographical situation, about its history for Europe.

Varese is a town of Lombardy, a door open to Europe. Varese has always been during centuries a crossing point of the merchants who exported the creative and artistic talent of Italian people and imported the concrete, industrious activity of other peoples of Europe.

Our people has sinthetized these qualities of creative talent and industry. Our town was one of the first that welcomed the European Communities experience.

In our town there is the European School, where the joung people learn how to live together.

In our town live many people, who are working in the Joint Research Centre of the European Communities at Ispra.

You have chosen as subject of your congress the solar energy for development and therefore, let me consider the cultural aspect of this subject.

Everybody knows how important the sun has been for all peoples through the history. Love and fear have men always experienced before the sun. Today this relation is still stronger. Today the knowledge leads the man to build his future with the use of solar energy. Today the man is discovering a new relation with nature, not only for necessity, but because he is discovering in nature his original nature.

Let me now, conclude by welcoming you in Varese, in the so-called "Garden Town" as built by its citizens who love the nature.

Today we want to defend the peculiarity of our town. In defending human values. We are sure that your presence in our town will be an useful example of common work for all peoples. And this is my wish for your congress.

Ladies and Gentlemen, welcome in Varese and thanks for that part of better world, that you are building in the coming days.

Thank you.

SUMMARY OF STATEMENT

BY

Mr. Maurice FOLEY,
Deputy Director General for Development
Commission of the European Communities

Two main barriers to development are chronic shortages of food and energy. The Community and the Commission are aware of this; their development efforts focus on these constraints. The Community fully participates in multilateral dialogues on these and related questions said Maurice Foley, Deputy Director General for Development at the Commission of the European Communities speaking at the Varese seminar on solar energy.

The Community spends about $1,200 million annually on project aid to overcome development constraints, notably food and energy. This is done through the Lomé Convention grouping 57 developing countries in cooperation with the European Community, through agreements with the Mediterranean countries, and an increasingly significant level of aid to the remaining developing countries. Large financial resources are available from Europe, including those destined to support regional schemes.

The developing countries are chronically short of traditional energy sources. For example, they depend for nearly 83% of their energy requirements on oil, largely imported.

Yet few requests for support from Europe in the energy field have been made. The developing countries need practical application of new energy techniques. This practical application at the level of the rural social unit does not require new resources; they are already available. The existing opportunities should be seized and rationalised. The new Convention to replace that of Lome, now being negotiated, should provide a continuing opportunity for the developing countries to utilise their own experience and that of Europe to maximalise this important natural asset. The Community stood ready, as this Conference demonstrated, to use the existing resources to the full for that purpose.

IV. Keynote Speech

SOLAR ENERGY FOR DEVELOPMENT

by

Sir George PORTER, F.R.S.
(The Royal Institution, 21 Albemarle Street, London W1X 4BS)

Commissioner Brunner, Mr. Minister, Mr. Mayor, Ladies and Gentlemen:

First, I should like to thank the Commission of the European Communities for inviting me to participate in this conference and to speak on 'Solar energy for development'. Anybody who, like myself, has spent the last few months in England can only maintain his interest in solar energy by taking a global outlook.

In his invitation to me to give this opening speech, Dr. Schuster asked me to be controversial. This I promised to do — but whom should I controvert? There are few people who take a neutral view about solar energy, there are the enthusiasts and the cynics and few between, although both extreme views are, like extreme views in most other matters, unreasonable. I will present both of them briefly.

1. The euphoric case

First, solar energy is natural. It is the source of life and its evolution, and it maintains life today. It is the basis of our technological development which could flourish only when abundant sources of fossilised solar energy, in the form of coal, became available. Over three billion years ago, photochemical reactions in the upper atmosphere began the synthesis of the simple organic molecules which were the building bricks of living matter. Two and a half billion years ago, when the atmosphere was becoming oxidising and therefore opaque to ultra-violet radiation, the first "green revolution" occurred. The green molecule, chlorophyll, became the primary receptor of solar energy for all purposes of life and remains so today.

The processes of life, evolution and technological development appear to be contrary to all our common experience because they are processes of apparently spontaneous increases in order whereas we know, and thermodynamics expresses this common knowledge precisely, that the natural state of things tends towards increased chaos or entropy. The paradox is, of course, easily resolved by noting that this law applies only to a closed system. Our world is not a closed system and it relies on an outside source of free energy or 'negative entropy'. That outside source is our sun.

Second, most of the energy we use today is solar energy. The sun maintains the temperature of the earth within reasonable limits which only need 'topping up' or, sometimes, a slight lowering, for greater comfort. It supplies light for most of the time that we are awake. It distills our water and raises it from sea level to the land. It provides all our food, and most of our clothing and basic building materials such as wood.

Thirdly, there is more than enough solar energy for all our needs. The solar energy falling on the earth's surface in twenty days is greater than all the fossil fuel reserves on earth. The present average energy demand of one person can be met, with 10% recovery, by an area 120 m² (11 metres square) between latitudes 40°N and S where 80% of world population lives and needs are greatest. An area of 600 km square (i.e. 360,000 square km) at 10% efficiency would supply the present demands of mankind. The projected ultimate demand, corresponding to a population of 10 billion, with an average standard of living comparable to developed countries (i.e. 5 × present, on average) requires 15 times this area, or 3 000 km square. This is 6% of the earth's land surface, one-eighth of which is desert.

Finally, sunlight is universally available to all people and most of it goes to those who need it most. Politicians, terrorists and trade unionists cannot stop the sun shining and it is pollution free — there is not even thermal pollution.

2. The Cynical (Realist?) Case

Everything on earth is free but it costs money to collect and use it. The sun's energy is very expensive to collect because it is (a) spread over large areas with low density, and (b) intermittent. It goes out when it is needed most.

All this adds up to one criticism only — the economic one. The low density can be overcome by covering large areas and the intermittency can be overcome by storage but these cost money.

It is a fact of life, which we must accept, that both individual people and countries as a whole, whether rich or poor, will try to get the best value for money. Value may be measured not merely in dollars per megawatt; it may include such things as security of supply, ease of maintenance, avoidance of pollution and care of the exchange money supply. But if, after taking all these matters into account, it is cheaper to get one's megawatts from a diesel engine than from a solar panel, the diesel engine will, and should be, what is used.

At present, in the *developed* countries, energy from fossil fuels is far cheaper than from solar energy. Even nuclear energy is cheaper, except for very special local applications. This will undoubtedly change in favour of solar energy as oil supplies begin to run out in the next few decades and coal becomes scarce in the next two or three hundred years. But, for the immediate future, solar energy cannot compete with this once-and-for-all supply and there is little evidence that anybody is prepared to save it for future generations.

So, for the developed countries anyway, the introduction of solar energy on a competitive basis will be very slow, if inevitable in the long term.

Are the developing countries different? The realist will say that, except for minor special cases, if a diesel engine is cheaper than a solar plant to install and run in a developed country, then it will also be cheaper in a developing country. If he is a real realist, he will recognise that nuclear and other large gigawatt installations are not applicable to countries without an electrical grid, but what are the relative economics of small fossil fuel plants and solar units? That is the essential question that we have to consider this week.

Developed and developing countries

It is interesting first to look back on the energy sources which man has employed over the last 100 years or so. Fig. 1 shows how the *proportion* of this energy has changed from being predominantly wood in 1850, to coal in 1900 to gas and oil today (the total energy use has, of course, increased enormously over this period).

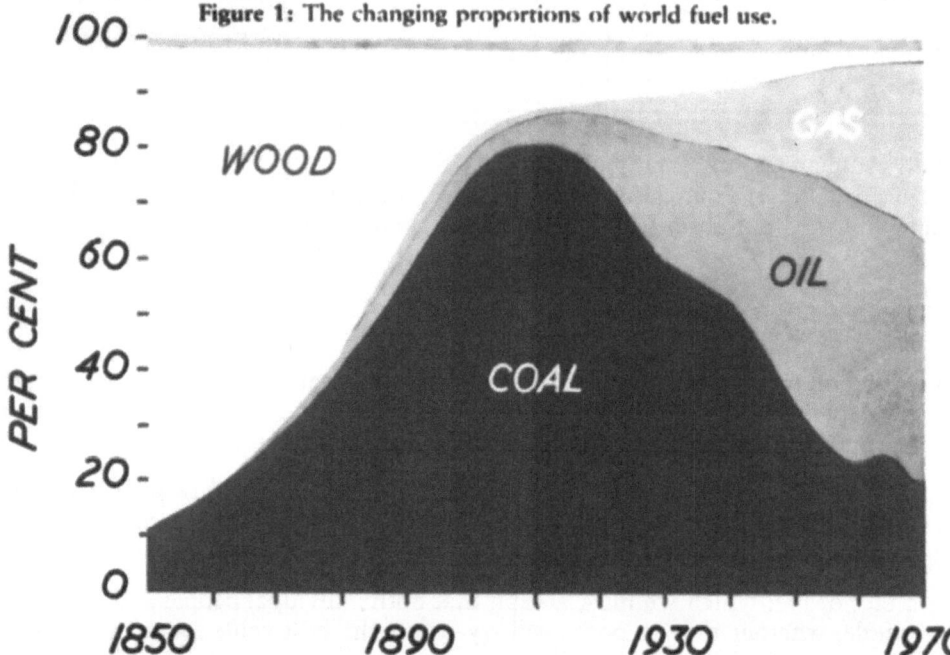

Figure 1: The changing proportions of world fuel use.

The sources of the world's primary energy in 1974 are shown[1] in Table 1, and we note that wood represents only 4.9% (although this may be an underestimate since the non-commercial wood collected by village communities is not accounted for).

TABLE 1
The Present Scale of World Energy Use

Source of Primary Energy	1974 World Consumption	
	10^{18} Joules	%
Coal	76	26.6
Petroleum	118	41.4
Natural Gas	48	16.8
Hydro and Geothermal	15	5.3
Nuclear	12	4.2
Wood	14	4.9
Dung	1.9	0.7
Crop residues	0.3	0.1
Total	285	100

However, when we distinguish between the developed and developing countries the situation is very different. As Table 2 shows,[2] wood and dung are of the greatest importance in the developing countries; they represent more than half of the energy use of India as a whole and are the only sources of energy in many villages whose total population, world wide, is of the order of one billion people. The energy used *per capita* by peoples in different stages of development is shown[1] in Table 3, along with the amount of land which would be needed to provide this from sunlight at 10% efficiency.

TABLE 2

Wood and Dung as % total fuel

Western Europe	0.7	India	56
World	6	Rural India	93
Latin America	20	Certain villages*	100
Africa	60		

* Approximately 1 million villages of several hundred or thousand people each, worldwide.

TABLE 3

Energy per head per year

Countries	Joules used ($\times 10^{-9}$)	Area required at 10% efficiency and 200 w/m^2
Developed	175	278
Developing	14	22
Poorest	2 - 5	3 - 8

The developing countries therefore already depend almost entirely on solar energy income. Two courses are open to us; we may improve and increase this income or turn to fossil (including nuclear) fuels. The latter will take time; can it be done significantly before fossil fuels become more expensive than solar? If not, it may be a waste of money and effort to carry out this form of development.

Whatever the source of energy provided, the technologically advanced countries could greatly help the developing ones without much sacrifice. A gift of 2% of their own energy consumption would double the energy of those in the poorest countries. The so-called developed countries consume 86% of the world's primary energy and their consumption per head is 13 times that of the developing countries. This is common ground whether it is acted on or not. The contentious point is, given this aid, what part should solar energy play now?

Allowing for food production, shelter and transport, it has been estimated that the minimum annual energy needed by a person in a country like India for a satisfactory life is 40GJ (1 450 kg coal). It may be reasonable, therefore, to consider about 10GJ increase above the present 14GJ as a reasonable short term goal. For the poorest sector, estimated at 1 billion people, this means 10^{19} J/year which is equal to the UK total consumption, 3.5% of that of the world, or 10% of our present annual use of oil.

The quality of solar radiation and its efficiency

Man needs his energy as low grade heat and as high grade energy in the forms of electricity, mechanical work and chemical free energy. These are interconvertible only within the laws of thermodynamics and to interpret these rules we must enquire about the quality, or special distribution, of the sun's energy as it is received on the earth's surface.

The sun is a nuclear fusion reactor whose temperature beneath the surface exceeds ten million degrees but whose surface temperature is about 6 000°K. However, its thermodynamic temperature at a flat collector or a leaf is only 1 350°K, largely because radiation is received only from the small solid angle subtended by the sun (6.8×10^{-5} steradians) whilst radiation is emitted from the absorbing surface over a whole sphere. Focussing collectors can increase the quality or temperature of this radiation, but not the quantity, by increasing the solid angle.

Without focussing collectors, the temperature of 1 350°K leads to a theoretical maximum efficiency of thermal conversion to free energy of 78%, though other losses reduce this efficiency in practice to much lower values. For a flat plate collector, producing a temperature elevation of 60°K, the theoretical maximum efficiency is only 18%. If we need only low grade heat for space and water heating or, in the developing countries, more particularly for crop drying, this can, of course, be collected with nearly 100% efficiency; but if we need high temperature heat of free energy we have to resort to very expensive mirror tracking systems or short-circuit the thermodynamic heat path by using 'quantal' absorbers.

'Quantal' absorbers result in the excitation of electrons in the material, either from the valence band to the conduction band of a semiconductor or from the ground state to an excited electronic state of a molecule.[3] Here, although the effective temperature is high and the corresponding thermodynamic limitation small, a second limitation on efficiency is imposed, which again is determined by the spectral distribution of solar radiation. The nature of the limitation is illustrated in Fig. 2. Usually a single material is used, which absorbs nearly all the solar radiation at wavelengths below the long wavelength "cut-off".

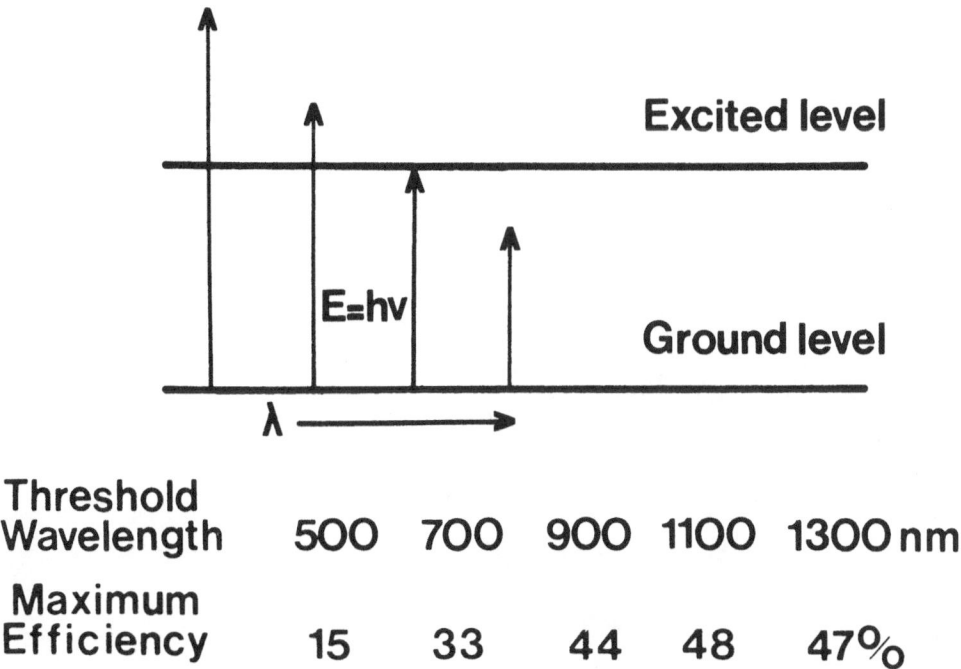

Threshold Wavelength	500	700	900	1100	1300 nm
Maximum Efficiency	15	33	44	48	47%

Figure 2: Maximum efficiency of solar energy conversion as a function of threshold wavelength.

At the cut-off wavelength the energy of the absorbed quantum is equal to that of the band-gap; wavelengths longer than this will not be utilised and quanta with shorter wavelengths have excess energy which is wasted. Using the distribution of solar energy for a clear sky at air-mass 1, the optimum wavelength cut-off for an absorber is seen to occur at 1 100 nm which happens to be that of silicon; this is one of the reasons why overall efficiencies as high as 17% are obtainable with this material.

The long-wavelength cut-off of chlorophyll — the principal absorbing material of green plants — is 700 nm and it seems that nature has, for once, been less than perfect in its design. But Landsberg has pointed out that in a cloudy sky, such as occurs frequently in Britain, the quality of solar radiation is different, the maximum being shifted to shorter wavelengths so that the optimum wavelength cut-off now matches rather well the absorption by chlorophyll. We can only conclude that God designed the green plants for the British!

The quantity of solar radiation and economics

The energy of solar radiation reaching the outer atmosphere of the earth is 1 353 w/m^2 and, after absorption by the atmosphere, this is reduced to an average of 800 w/m^2 continuous radiation over a circle of area πr^2, where r is the radius of the earth. Since this has to be distributed over the surface area $4\pi r^2$ of the globe, the energy received, averaged over day and night, winter and summer, equator and poles, is 200 w/m^2. Certain desert areas receive 300 w/m^2 and countries in middle Europe receive about half of this. These points are illustrated in Fig. 3.

Figure 3: Energy of solar radiation reaching the surface of the earth (circle) and its distribution averaged over the year (globe). The smaller white square on the base of the globe is the area required to supply, at 10% efficiency, the present energy consumption of mankind.

A simple calculation of the costs of collecting solar energy can be made on the basis of these figures. Let us consider an average insolation of 200 w/m^2 and an efficiency of collection of 10% (i.e. 20 w/m^2 is collected). A typical figure for the capital cost of a large conventional or nuclear power station is $1/watt of installed capacity. To be competitive with this our solar collector must cost $20/m^2$ or less.

In isolated areas, and for special purposes, power is more highly valued and solar collectors may be competitive at much higher costs. A few other costs of power and collector are compared in Table 4. Even the highest of these figures, $200/m^2$, is considerably less than the cost of silicon photovoltaics or of a mirror-concentrating thermal system with tracking. Costs of photovoltaics are falling

TABLE 4

Solar collector at 200 w/m² and 10% efficiency

$/m² of collector	$/watt of capacity
10	0.5
20	1
40	2
100	5
200	10

rapidly but encapsulation forms a major part of them and they are unlikely to fall to $20/m². It is difficult to see how high temperature thermal systems with mirror tracking can ever be of use, especially since they are large scale structures which would have to compete with conventional power stations. Small mirror concentrators are, in principle, applicable to cooking purposes but the art of, for example, the Indian cook depends on heat processes quite different — and more reliable — than overhead radiant heating. Photovoltaics are certainly very useful and economical for small local supplies of electricity such as are needed for communication and television in remote areas.

There is a further item of cost not yet included but usually essential for electrical generators-storage. In cases such as water pumping intrinsic storage is provided but for most purposes chemical storage batteries must be used which will add more than $25/m² to the above costs.

Storage and transport of energy are of vital importance in all these costing exercises. Electricity cannot be stored, heat and potential energy can only be stored for limited periods and, for inexpensive long term energy storage, we must always resort to chemical potential. For this reason, conversion of solar energy into chemical fuels or biomass is the ideal way to overcome the intermittency of the source — as nature discovered long ago.

Photosynthesis and biomass

Nearly all the energy which we use today has been collected and stored by this route and I believe it offers the most promise, of all solar energy methods, especially for developing countries. There is often a tendency to identify power needs with electrical power whereas, even in the developed countries, it represents only about one third of our energy use. It is the form of energy supplied by longer term facilities such as coal, nuclear and hydroelectric generators whereas the more immediate crisis which confronts us is a shortage of gaseous and liquid chemical fuels. Finally, for the developing countries, biomass is the only form of solar energy which is effectively used at present and which can be further developed without great capital expenditure.

Energy is stored as biomass by the reaction

$$CO_2 + H_2O = (CH_2O) + O_2 \qquad \Delta G_o = 470 \text{ kJ/M}$$

This involves the transfer of four electrons but it is well established that each electron transfer requires two photons so that eight photons in all are required.

$$2H_2O \xrightarrow{4h\nu} O_2 + 4e^- + 4H^+$$
$$CO_2 + 4H^+ + 4e^- \xrightarrow{4h\nu} (CH_2O) + H_2O$$

At the long wavelength cut-off this means an energy efficiency of the photosynthetic reaction of 33% and, of course, a reduced efficiency at lower wavelengths. Using our previous figure for clear sunlight and a 700 nm cut-off (33%) we arrive at a maximum efficiency of 11%. We also have losses due to reflection by the leaf and by respiration giving an expected overall efficiency of 5-6%. This is the best efficiency which is actually found in practice but under field conditions the efficiency is lower. Some typical yields and efficiencies of crops under peak growth conditions[4] are given in Table 5. Table 6 shows how these are reduced in practice.

TABLE 5

Maximum Growth Rates of Experimental Crops

Crop	Growth rate $(g.m.^{-2} day^{-1})$	Efficiency (% utilization of solar radiation)
C_4-species		
Bulrush millet	54	4.2
Maize	52	4.6
Sugarcane	37	3.7
C_3-species		
Rice	36	3.2
Sugar beet	31	4.5
Soybean	27	4.4
Pine (Pinus radiata)	41	2.7

TABLE 6

Efficiency of Photosynthesis

Field conditions (highest at peak growth)	2-4.5 %
Field conditions (highest year-round)	1-2 %
Average over globe	0.15%

In spite of the low efficiency this is a tremendous amount of stored energy. It is equivalent to 2.7-3.5 $\times$ 10^{21} J/annum which is ten times the present world consumption of primary energy.[6]

Taking the energy of dry biomass as 1.75 $\times$ 10^{10} J/tonne and productivity of 30 tonne/hectare for the intermediate typical case of sugar beet, we obtain the areas given in Table 7 necessary to supply one person with his present energy consumption.

TABLE 7

Energy Usage and Equivalent Area at 30 tonne/hectare

Countries	Energy/person/year	Area required
Developed	175 GJ	0.33 hectare
Developing	14 GJ	0.025 hectare
Poorest	3.5 GJ	0.006 hectare

N.B. 1 hectare = $10^4 m^2$.

Looking to the future, I would like to mention briefly the promise of purely photochemical methods for the storage of solar energy in chemical form. Apart from the low efficiency of living plants they demand large quantities of water and of fertiliser if they are to grow with the efficiencies given in Tables 5 and 6. Nor do they give us the products we need and further processing is necessary — often of a quite complex and expensive kind. Research into *in vitro* methods of producing hydrogen from the photolysis of water and organic liquids from water and carbon dioxide is progressing quite rapidly[5] and, if the problem can be solved in the laboratory, its development on a large scale could be a relatively rapid process — the lead times need not be as long as we are accustomed to in other fields.

In conclusion, my own feeling is that research and development is still greatly needed in almost every area of solar energy, whether it is for developed or developing countries and this research must, on the whole, for the time being anyway, be done in the advanced laboratories of the developed countries. They have an interest in this quite as great as the developing countries. The coming fuel shortages will be even more traumatic in countries like the United States, whose whole life would grind to a halt without huge quantities of power, than in the least developed countries which would at least know how to survive.

Over the next few days the many ways in which existing solar technology might be more effectively applied in the developing countries will be discussed. We must not, if we are to keep to the point, discuss the importance of water pumps, refrigeration etc., which is wellknown, but the relative advantages of solar over other forms of energy as power for these devices. The unique advantage of solar energy in these countries lies in the isolation of their villages so that small amounts of power are far more highly valued than in the developed countries.

Of all the ways of collecting and storing solar energy, the one which, it seems to me, might be most useful to these countries now is agriculture and the growth of energy fuels by agricultural methods. These are labour intensive, of course, but this is not necessarily a serious disadvantage in countries where labour is plentiful. Suitable crops for this purpose have not been very intensively researched, even amongst existing species, and many others may be developed by selective breeding and even genetic engineering. The economical use of water by plastic greenhouse structures is also promising.

I should like to finish on an optimistic note, even though it is a longterm view. Today, advances in technology can only occur by international collaboration. No country is self-sufficient in minerals, in knowledge and technological knowhow, in energy sources and food. On the whole, the less developed countries are rich in solar energy and their land is often poor. These may well become the energy exporting areas of the future. For example, the poorest people are surely those who live in barren deserts with no oil beneath them. It will be far more economical to grow the world's chemical fuels in these areas and ship them overseas than to grow them on good land which is needed for food production. If the technology for this can be developed, these areas will become the power houses of the world.

One eighth of the earth's land surface is desert with little potential other than its sunshine. If half of this were used to provide stored energy as chemical fuels, with 10% efficiency, it would supply a population of the world of 10 billion (the hopefully projected maximum population) at a standard of living in energy terms, similar to that of the average European. This is surely all that can ever be required. But this ultimate goal, which may or may not lead to the maximum happiness of the greatest number, reminds me of some words of Linus Pauling last year. He said that more technology, at least in the advanced countries, is not likely to increase our happiness and that instead of trying furthert to **increase happiness,** our technology should have, as its ambition, the **decrease of unhappiness** in the world. This can best be done, and must most necessarily be done, in the poorest countries where real unhappiness is readily alleviated by a few kilowatts of power. That is the challenge which is before us in the discussions of this week.

References

(1) *Renewable energy sources for developing countries,* Report from Energy Technology Support Unit, AERE Harwell (1978).

(2) Working document of this meeting.

(3) G. Porter and M.D. Archer, *Interdisciplinary Science Reviews,* 1, 119 (1976).

(4) N.K. Boardman, Royal Society Discussion on Solar Energy, *Proc. Roy. Soc.,* 1979.

(5) G. Porter, *Proc. Roy. Soc. A362,* 281 (1978).

(6) D.O. Hall, *Fuel,* 57, 322 (1978).

V. Water

A. KEYNOTE SPEECH

by

F.J. Friedrich
Project Management
Non-nuclear Energy Research
Jülich, Federal Republic of Germany

Ladies and Gentlemen,

this introduction to water processing by solar energy has two aims:

— It will provide a general introduction into the topics of solar water pumping, water desalination and closed system agriculture.

— The second aim is to raise additional questions in order to intensify the dialogue between experts and decision makers. In this way I have been asked by the organizers of the conference to make some provocative statements in order to stimulate the discussions.

I highly appreciate the fact that I have been invited to discuss this combination of topics: Sun and water. Both are the necessary prerequisites for life on this planet. It seems interesting to state that sunlight and water have several common properties:

— both are available on this planet in great abundance

— however man has to accommodate himself to their peculiar properties and geographical distribution

— finally, both can be mortal to man if he does not react adequately.

The hostility of a desert region to life is quite apparent. Even if the human being is inactive, the extremely low humidity will desiccate his body in a few hours if drinking water is not available.

WATER PUMPING

It is relatively simple and has been standard technology for more than 4 000 years to supply water by means of ditches or stony channels even over large distances and in excessive quantities.

Even if dams are required this is in most cases the preferable method for supplying water.

Water pumping is necessary whenever the aforementioned method cannot be applied. This may be true for distant or geographically isolated oases where the artesian well is insufficient or exhausted. Water must then be pumped from the ground water table which can be at any depth. The necessary mechanical power can be supplied by men, animals, diesel engines, biogas, by the wind or by the sun.

Let us now discuss these options.

The proposal to use manpower for water pumping can only be submitted by a European who has never performed hard work in a tropical country. I propose that we refuse this option.

Animal power is certainly worth some consideration. We scientists tend to be fascinated by the most sophisticated technology like nuclear reactors, stirling engines and photocells, and we often neglect simple technologies.

But we have to evaluate all possible options! Certainly there are power limitations and the character and the availability of animals and their living conditions have to be considered. In fact, these simple and cheap systems are hard to beat by modern technology.

Biogas motor pumps have not yet received adequate attention. Wood gasifiers, for example, are simple and easily servicable. Many thousands were used on trucks and private cars in Germany during World War II. Where wood is not available, pyrolysis or fermentation of agricultural waste may be considered.

The use of wind energy for water pumping is definitely the most preferable option if climatic conditions justify the installation of windmills. Certain maintenance problems have been reported. Nevertheless hundreds of thousands of small windmills have been in operation around the world.

Solar thermal water pumps

Finally sun powered water pumps are the most modern option. I do not hesitate to praise Girardier as one of the most important pioneers in solar energy utilization, especially in view of the fact that he introduced his pumps 15 years ago.

There are two main reasons why this system has not yet found widespread application: high installation cost and maintenance problems.

Let us first consider cost aspects. For this discussion we rely on the available information on solar thermal power stations.

TABLE 1

Solar-thermal-mechanical energy conversion (1), (2)

Type/Location	Output	$/kW	Status
Pump/Kenya	1 kW water	50 000	operation
MBB/India	10 kW$_{el}$	100 000 *)	operation
DS/Egypt	10 kW$_{el}$	200 000 *)	operation
AEG/Greece	10 kW$_{el}$	50 000	construction
MAN/Spain	40 kW$_{el}$	50 000	construction
IEA/Spain	500 kW$_{el}$	40 000	design
EEC/Italy	1 MW$_{el}$	20 000	construction
Emergency Generator	1 kW$_{el}$	500	mass-production
Diesel pump/India	4 kW water	150	mass-production
Coal fired station	600 MW$_{el}$	400	typical cost

*) Include training programmes.

46

This is permissible since the solar part is quite similar and the cost of the electrical systems are well known. There exists a certain cost degression if the output is increased. The 1 kW pump fits correctly into this line. It is simpler but also smaller. We have a long way to go considering the competitive conventional systems listed at the bottom of the table.

The following question has frequently been heard: "Why can't we produce a solar pump like an automobile?" — One kg of automobile costs 7 $ in Europe and $ 5 in the USA. This is in fact an incredibly low price for a high concentration of sophisticated technology! The components for a solar pump can certainly be produced even more cheaply ex factory. Why not now?

The answer is that the low automobile price is the result of 50 years of continuous refinement in mass production. Several hundred millions of cars have been built. Imagine the number of scientists and engineers who have introduced their ideas! — Considering all these arguments I have little hope for a cheap solar thermal pump in the near future.

Certain maintenance problems connected with this type of pump have been reported. I think they are not too serious. Such problems are always connected with the introduction of a new technology and can be overcome by adequate training of the operator and continuous feedback of maintainance reports to the manufacturer. Each new car model also has its growing pains.

Photovoltaic pumps

Today photocell pumps in the 1 kW power range are favoured. I think this is justified. Maintenance problems have not been reported. Two drive systems are in use. One employs a dc motor and a long pump shaft to the underground impeller. The dc motor is a new development with a good initial breakaway torque required for the start up in the morning. The other system employs a three phase semiconductor inverter which is matched to the asynchronous motor. This solution is preferred by countries which already have a production line for ac motors. However some energy is lost in the inverter.

Future prospects for photovoltaic pumps are exceptionally favourable. They fit perfectly in the present technology development pattern, which may be characterised as follows:

— the trend to substitute mechanical systems by electronics and
— the tendency to reduce the number of installation work hours at the building site by preassembling as many items as possible in the factory.

Solar thermal or photovoltaic pumps?

The above trends are a reaction to the increasing labour cost in the highly industrialized countries. But is this also valid for developing countries? Most of them do not have to face the problem of high and continuously increasing labour cost. Theoretically it could be advantageous there to install solar thermal pumps in large numbers. There should be sufficient manpower to perform the high amount of labour which is connected with the installation of these systems and there should be workers to operate and service the pumps.

These arguments lead to the following provocative questions, which frame a wide range of intermediate solutions:

— is it more advantageous to import photocells forever or to buy an automated production line for their manufacture, or

— is it more advantageous to build up an independent national solar thermal industry providing labour for millions?

I think each country has to find its own answer.

Water requirements

The absolute minimum for the survival of a man in a hot arid region is 3 l per day. Under this condition he can neither move nor work. From various statistics I quote these standard requirements:

Men	20 l per day
Sheep	15 l per day
Cows	80 l per day

If irrigation for agriculture is required the water consumption rises excessively.

TABLE 2

Typical water demand for various irrigation systems at 25° North, average values in m³ per day and per ha

Water demand	January	February	July	August
Consumptive use of plants	14	19	55	48
Water demand with				
simple surface flooding	20	26	75	64
sprinkling/spraying	17	23	67	57
drip irrigation	16	21	62	53

As a general rule, for the aggregation of 1 kg of dry plant substance 300-500 kg of water are required. This is, by the way, one serious argument against energy crops in arid regions. The water demand for irrigation depends on the system. Improved water economy is paid for by increased system sophistication.

WATER DESALINATION

Potable water must have a relatively low salt content. There exists no precise definition for brackish or salty water (3).

TABLE 3

Water classification

Type of water	Salt content p.p.m.
Drinking water	
desirable	max. 500
permitted (USA)	max. 1 000
tolerable	max. 1 500
Brackish water	1 500-10 000
salt water	10 000
sea water	
Oceans	34 000
Arabian Gulf	43 000

To a certain extent brackish water can be used for irrigation. But this requires careful investigation in order to prevent salt accumulation in the soil. Additional water is required for leaching and a draining system may even become necessary.

In general water desalination may be considered for:

— supply of drinking water

— irrigation of top quality products

— blending of brackish water for irrigation.

Water desalination is definitely required for the supply of drinking water to men and cattle, if the water table provides either no water or an insufficient amount of good water. This requirement has absolute priority. Other applications are rather questionable.

Water desalination is a standard technology and the economical conditions are well established for large installations which may have outputs of more than 100 000 m^3 per day.

More than 80% of the total amount of desalted water in the world are generated by evaporation systems which are generally coupled to fossil power stations.

TABLE 4

Energy demand for desalination (4)

Type of process	Input water	Energy required for 1 m³ of desalted water
Evapration	seawater	50-250 kWh thermal
Electrodialysis	brackish water	3-7 kWh electric
Electrodialysis	seawater	17 kWh electric
Reverse osmosis	brackish water	2-3 kWh electric
Reverse osmosis	seawater	15 kWh electric

In general the energy demand for desalination is quite considerable. In addition, small units have poor efficiencies.

Solar energy evaporators have been used now for 100 years. The output of the simple greenhouse evaporators is below 3 l per m² per day. More sophisticated systems with heat recuperation reach 5 l per m² per day. Multistage units are being designed with daily outputs of 10-20 m³. They will reach efficiencies of 15-20 l per m² per day (5). The multistage units require some electrical energy for the vacuum pump and other ancilliary pumps and controls. This amounts in general to approximately 10% of the thermal energy.

Reverse osmosis is gaining importance especially in view of the fact that the energy requirement can be matched to the input water quality. Small reverse osmosis units are under development and these use photocells for the energy supply of the pressure pump.

The more efficient desalination systems require skilful maintenance and in some cases water pretreatment. Reliable experimental data are not yet available. This is why multistage and osmosis systems will be tested in Mexico under comparable conditions.

Considering the energy demand it is completely uneconomical to consider desalination for irrigation with present solar energy technology. The desalted water is so expensive that it must be reserved as drinking water. It is generally more economical to conduct irrigation water over large distances even if pumps have to be used.

Closed system agriculture

Elaborate agricultural technologies have to be applied in regions where good natural water cannot be obtained. A variety of solutions have been proposed, all based on the same principle: the recollection of the excessive amount of water transpirated by the plants (Table 2) by using an air tight containment.

In one system which is operating in the Atacama desert in Chile, brackish water of any quality is trickled over the outside surface of the platic greenhouse roof. Condensation takes place on the inner surface of the roof and the condensate is recycled to the roots of the plants. Water losses have to be replaced by an independent desalination unit.

Another system uses a conventional glass roof and has a built in desalination system. It is much less sensitive to water losses (6).

Both systems require careful maintenance and skilful operation.

CONCLUSIONS

1. Solar energy has a low power density, as we all know. It follows that the use of power consuming devices or systems should be avoided, as for instance desalination.
2. If high power is needed large structures have to be built. They are costly and require servicing.

For the reduction of cost new ideas and high level manufacturing experience are necessary. For reliable servicing sufficient training and the readiness to shoulder responsibility are required.

All these requirements can be accumulated under the heading: Education and Training.

New Ideas

It is my personal experience that there are a great number of excellent scientists in many tropical universities. They have brilliant ideas, but many of them are engaged in theoretical work. This is one reason why technological progress is much slower than it could be. But there are technologists and applied physicists and we hope that their number will increase. Such men are needed for the introduction of solar energy technologies. The situation could be much improved if at least introductory courses on solar energy would become mandatory for students of the engineering sciences.

Construction and maintenance

Construction and servicing problems are dependent on medium level education. In the industrial countries there exists a literacy pyramid: as an approximation the number of people is inversely proportional to their education.

This pattern has a long tradition. But there exists a gap in this pyramid in tropical countries and this gap has to be filled. For solar installations, workers who can solder, weld, make threads, cut glass and gaskets etc. are required. Foremen are needed wit additional abilities and technicians who can perform simple computations etc. Such people are only available in small numbers if at all. Several countries have already reacted accordingly and have installed schools for technicians, but so far not too many.

I am not able to propose a solution for the following problem: Imagine an experienced technician operating and servicing a solar thermal power station with desalination in a remote desert location. How long will he stay there, knowing that he can have a well paid job in a major city?

All problems cannot be solved at once. But they must at least be identified and have endeavoured to point my finger at some major ones. I hope I have been provocative enough.

References

(1) Annual Report 1977 on New Sources of Energy, Project Management for Energy Research, KFA-Jülich, P.O. Box 1913, D-5170 Jülich.

(2) Parikh, SOLAR ENERGY, Vol. 21, p. 103.

(3) Michael d'Orival: Water Desalting and Nuclear Energy, Thiemig-Verlag, München, 1967.

(4) H.K.J. Hauser, UMSCHAU, 1977, Heft 9, p. 268.

(5) Private communication, G. Lippmann, DORNIER-SYSTEMS, Friedrichshafen, Germany.

(6) R. Bettaque, ZEITSCHRIFT FÜR BEWÄSSERUNGSWIRTSCHAFT, 1977, Heft 2, pp. 3-24.

B. SUMMARY OF THE SESSION

by

P. LEQUEUX, General Rapporteur
P. DUNN, B. DEVIN, A. STREHLER, Rapporteurs

Water is indispensable for all forms of biological life, and naturally it plays a large part in the fundamental needs of man. While it might be considered that the expression of Leonardo da Vinci that water is the motor of nature is an over statement, it is nonetheless true that water and the provision of water is essential for development. If it is abundant and easily available, water can facilitate all forms of development; if it is difficult to obtain or obtained only at great expense, settlement is difficult and desertification a constant threat. This was the case during the drought in the Sahel 1972-1974. As the water cycle is a closed one, there is probably enough water in the world to meet all future needs. Unfortunately water resources are badly distributed geographically and seasonally and often vary in quality.

In order to respond to the needs of developing countries, industrialized countries are improving their techniques and methods to increase the efficiency of water supply, bearing in mind the natural resources and finances which are available. Developing countries have an interest to develop water resources in the best conditions within an economic, social, technical and environmental framework, which will vary and be different from the needs of industrialized countries.

A large number of water problems can be resolved by the exploitation of solar energy. How far can solar energy bring an efficient and economic solution for water problems?

Solar energy is to-day playing a role in five key sectors in the economy of developing countries, namely human hydraulics, pastoral hydraulics, cleaning, industrial hydraulics and agriculture. Two types of principal application, pumping and desalinisation of water, have been identified. A third application concerns water economy and conservation.

This problem is important, because access to water resources implies the construction of barrages and pumping stations, for which the investment cost is considerable. It is necessary then to pay great attention to the problem of water management. One of the ways to obtain a productive use of water resources is by solar green houses in agriculture.

Energy for water pumping is a major requirement in the rural areas of developing countries. The power level required is generally 5 kW and normally around 1 kW for an individual installation, though there are some applications at higher power levels. Already such equipment has been the object of numerous applications throughout the world and the results appear very satisfactory. These applications are in the field of human hydraulics, pastoral hydraulics and agriculture. The use of solar energy in human hydraulics is above all seen at the village level, but it can also be applied in urban communities. Pastoral hydraulics includes those problems relating to two forms of cattle rearing — nomadic along cattle tracks and sedentary in ranches and in the villages.

Finally **agriculture constitutes**, with village hydraulics and pastoral hydraulics, the third sector **where solar energy can be used for pumping**. Further development is merited.

The **diffuse distribution of rural population suits the development of pumping by solar energy. Solar energy is also well adapted to the rural world**, where consumption is still relatively low. For example, a 1.8 kW pump will bring from a depth of 40 meters almost 60 m³ of water a day, sufficient for the needs of a village of 500 people, who could cultivate some hundreds of acres of irrigated crops and keep approximately one hundred head of cattle. Such equipment is already economically profitable, and is meeting with increasing success.

On the other hand, the power necessary to irrigate large surfaces involves high investment costs and very sophisticated equipment.

1. WATER PUMPING

Water pumping with solar energy conversion methods can be classified under direct conversion and indirect conversion.

1.1. Direct Solar Conversion

Most work on heat engine driven pumps has been carried out using the Rankine cycle. A number of 1 kW sets have been installed in the sub-Saharan regions, Mexico and elsewhere. Typical performance figures are 5 m³/hr of water from a depth of 20 m at an initial capital cost of 90,000 US$. Some engines of higher power up to 50 kW have also been operated and others are under construction. Some maintenance troubles have been reported.

There is some interest in liquid piston pumps and programmes are in hand in several countries including India and Papua New Guinea. This work is still at an early stage of development.

A number of electrically driven pumps powered by photovoltaic cells at around 1 kW pk are in operation in various regions, including West Africa. Though expensive, these pumping sets are reliable, largely maintenance free, light in weight and offer the possibility of significant cost reduction in the future.

1.2. Indirect Solar Conversion

Methane generation from biomass was referred to and its use either with conventional internal combustion engines or with liquid piston internal combustion engines was discussed briefly.

Wind driven pumps employing conventional horizontal axis mills are in use extensively in some regions. This use could be extended. There are also a number of new designs which may offer advantages over the conventional machines.

Hydro Power is highly site-specific but has applications in suitable areas. It was referred to only briefly in the discussions.

1.3. Selection Criteria

In addition to technical requirements, economic factors and social acceptability are of great importance in selecting a pumping system. The importance of cost and the desirability of local manufacture, leading to readily available maintenance, were emphasized in the discussions.

There is a general need for more field testing, including social and economic tests. Failures should be reported since these may be more informative than partial successes.

The size and technical level of the countries represented varies widely. For example India currently produces 10 kW (e) per year of solar cells whereas Mali has little industrial infrastructure. For such reasons it is not possible to give general solutions. However, the solar pumping methods can be roughly classified by stage of development.

a) *Established designs:* Windmills.

   Further work is required to produce a design suitable for manufacture in new regions and to implement its use.

b) *Technology available:* Methane generators to power conventional engines. The technology is established but further field work is necessary on socio-economic problems and implementation.

c) *Prototype development:* Heat engines, photovoltaics, vertical axis windmills. Further field assessment of socio-economic nature required.

d) *New ideas:* Liquid piston heat engines. Further research and development required.

2. DESALINATION OF WATER

Thanks to the aid given by the public authorities for research and development in the field of the production of drinking water the technology of desalination has become, in a majority of industrialized countries, a successful technique.

The choice of desalination procedure which is most appropriate is a function above all of the quality of the water which is locally available. Another factor in this choice which is just as important as the quality of water is the energy requirement. In the production of drinking water from sea water or brackish water, energy consumption is always relatively high and in large installations which use the evaporation principle, energy consumption is a major cost factor, which can be as high as the labour cost. For this reason the reduction of energy consumption is at the centre of all R & D activities in the desalination sector.

Therefore, solar energy represents a major benefit in the development of these techniques. It has been used for a long time as an energy source for the production by distillation of drinking water, starting from brackish water. Solar energy evaporation installations constitute often the only possibility for producing drinking water for the supply of small towns, for hotels, hospitals, schools and for rural settlements. They can also provide the distilled water necessary for the electrical batteries which are in use throughout the developing countries. In numerous sunny regions, generally poor in drinking water and in conventional energy sources, solar desalination has particular importance and is often much superior to other methods of desalination.

The conventional type of distillation procedure (the glass house), which is widely used in the developing countries, is not very efficient because of the large space it needs and the poor use of heat. On average, such equipment can produce 1,000 litres of drinking water a year per square meter of collector surface. These systems, which are relatively small in capacity, are already being constructed in several developing countries.

Besides solar stills, which are in operation in some villages in various countries, new experiments on more sophisticated desalination devices are under way in co-operative programmes. As an example, a 500 m^3/day multiflash unit studied in the Netherlands for water supply to an island was reported. The upper temperature of this unit is suitable for flat-plate collector operation. Mexico quoted comparative testing of multistage and reverse osmosis equipment (photovoltaic) on sea water in Baja California in cooperation with F.R.G. France reported details of the projected installation of a 60 m^3/day reverse osmosis unit powered by a thermodynamic solar engine for the Egyptian Government.

A specific problem of water purification was presented by Bolivia. River pollution with floating effluents from the tin mining separation process is harmful for many villages and lakes and decontamination is required for human consumption. No electricity is available along the river and the Government is seriously concerned with the problem.

A similar case was reported in Mexico with arsenic and boron contaminants. Both situations could benefit from membrane separation techniques, at the consumption level or preferably at the pollution site.

It was proposed to combine salt production and fresh water production in solar units, but prospects are not favourable due to low water yields in the final stages of salt production.

LIBERIA and SIERRA LEONE drew attention to the necessity of the local production of solar equipment, thus serving two purposes — job creation and satisfaction of a need. It was considered important to include the local manufacture of basic components: plates, tubes, etc.

Experts from Jordan pointed to the misjudgement which could be made by crude comparison of apparently correct economic estimates. Detailed comprehensive analysis of the specific water problem in its environment must be carried out in order to avoid the unwarranted rejection of new technology by hasty evaluations. The example of cattle herds in Jordan which receive water by expensive long distance trucking, was given.

3. GREEN HOUSES FOR WATER ECONOMIES

In dry regions pumping or desalination of water are costly operations which justify the implementation of techniques to economise water. It is in this perspective that the problem of green houses was presented at this conference, particularly with reference to agriculture.

The first way to economise water in arid zones consists of cultivating under plastic during periods when the temperature is low and evaporation minimal. Water economy can also be achieved by the use of appropriate technologies e.g. composting and tilling of the soil to promote root formation, natural or artificial shadowing, wind breaks and the selection of crops resistant to draught.

The green house is a covering which is permeable to the sun's rays and reduces transpiration at the price of incrasing the temperature of the plants. There is a difficult compromise between high consumption of water and excessive temperature during the heat of the day that must be found.

A technically well-developed solution is to utilize a selective filter to intercept infrared radiation from the sun. A heat transfer fluid circulates in the storage system (water or stones). The stored heat is used to limit the fall in temperature during the night. This method is particularly well adapted for the desert zones, in which it is necessary to protect against a strong evaporation and transpiration during the day and against a pronounced nocturnal cooling. It is as yet costly (50 $US per m^2 as compared with $ 5 per m^2 for the cheapest plastic shelters). It reduces water consumption by half.

Another solution consists of using a closed system embodying a sufficiently strong cold source (evaporation of brackish water or a nocturnally-cooled store). These completely closed systems can lead to a consumption of water which is extremely low. They require maintenance of the concentration of carbon-dioxide (which diminishes naturally in these conditions because of photosynthesis) and the control of parasites, which develop easily because of the high humidity.

4. GENERAL ASPECTS

Certain general aspects are found in all three technologies, and should be commented upon. Thus we have discussed in the majority of cases that there exists a relationship between the solar data, the wind data and the availability and quality of water. In research and development, it appears imperative to improve performance, develop new systems, draw up industrial standards and develop modular equipment with the aim of reducing investment costs while at the same time maintaining the simplicity, durability and profitability of such equipment.

A fundamental idea to retain in the promotion of solar energy is the adaptation of the technology to the conditions of the developing countries. All systems and equipments which are designed in the companies of industrialized countries, which do not bear in mind the technical environment, social conditions, economic possibilities and the finances available to the developing country for which the equipment is destined, are doomed to failure. Such adaptation can only be undertaken with the participation of the developing country, which must make at all level a contribution to the new technology.

Other aspects, such as the industrialization and the commercialisation of the equipment, transport, installation on site and the training of the staff must be borne in mind.

Another point is that the installations must be monitored to check their performance, maintenance, the quality of after-sales service, and their acceptance by local administration and population.

Finally, this is the opportunity to insist on the importance, for each country, of a solar energy policy with medium and long term objectives in selected areas.

VI. Power production

A. KEYNOTE SPEECH

BY

Henry DURAND
Président
Commissariat à l'Energie Solaire
Paris, France

INTRODUCTION

The unequal distribution of energy production and consumption among the major economic zones of the world is too well known to leave any doubt that availability of energy and the level of economic development are closely linked:

	Production	Consumption
OECD	40%	63%
OPEC	30%	4%
East European Communist countries	24%	23%
Developing countries (excluding OPEC)	6%	10%

These figures have to be corrected by the contributions from "non-commercial" energy sources, in particular biomass, to which I shall return later. Is it conceivable that this inequality could be adjusted by utilization of solar energy which is distributed abundantly over a large part of our planet, especially over the inter-tropical zone in which most of the developing countries are located? To find an answer to this question is the prime objective of this Conference and I shall endeavour to make my own contribution with regard to the production of electromechanical energy.

Among the various forms of energy consumed this is the one that is often regarded as the key to development, and that with good reason for it is the noblest form of energy, ranking first in the energy hierarchy.

According to the second law of thermodynamics, mechanical and electrical energy are two variants that can be converted one to another with a high degree of effeciency. I will run briefly over the various methods of producing electromechanical energy directly or indirectly from sunlight; for this purpose a short list will suffice, in which my classification is slightly different from the usual one:

Use of the thermal form of solar energy:

Low-temperature thermodynamic stations (flat-plate collectors);
Medium-temperature thermodynamic stations (concentration);
Wind power;
Hydroelectric power.

Use of the quantum form of solar energy:

Photovoltaic;
Biomass.

Just a remark in passing: of all the known forms of energy, photovoltaic conversion is the only one that produces electricity directly without any thermal intermediary. All the other methods that are known and used pass through a thermal cycle: an artificial cycle in the case of the fossil, nuclear, geothermal and vegetable sources or the thermal gradient of the seas; a natural one in the case of wind, wave and hydroelectric power. The purists would no doubt wish to make an exception of tidal power, which is a mechanical/mechanical conversion of the gravitational energy of the earth, and of fuel cell energy, which is a non-thermal conversion of the chemical/electrical type.

In this report I shall be mainly concerned with power generators of a fair size, capable of supplying a small built-up area in an isolated situation. I shall therefore refrain from mentioning the small photovoltaic generators of less than a few kW capacity, which are so widely used today in telecommunications or rural pumping systems.

The obstacle to the introduction of the solar production of electromechanical energy in the economy of the developing countries.

It needs hardly be emphasized that among the developing countries there is wide diversity, some of them having already embarked on vigorous industrial activity, whilst others are still wholly dependent on imports for all manufactured equipment. Under these circumstances, it is hard to make other than very general comments, some of which will be more applicable to the situation in one country and some to another.

To my mind, there are three types of difficulties that hinder the penetration of electrical solar energy in the developing countries:

a) One is technical: owing to the vagaries of the climate, there is no assurance that energy will be available when required; this is a problem that can be solved only by storage on a substantial scale. In this respect, biomass is an exception in that it is the only source of energy in the list given above that is both renewable (at any rate in an annual cycle) and storable;

b) the second is economic: solar energy, especially in its electromechanical forms, requires a large initial investment which can be obtained only by international cooperation, i.e. by the transfer to the developing countries of part of the capacity of the industrialized countries to provide permanent capital. Here too, biomass is an exception to the rule since the fermentation or gasification equipment that accompanies the gas-driven generating plants is a fairly modest additional requirement;

c) the third is technico-economic or social: these are fairly new technologies which have not reached full development even in the industrial countries. They are not necessarily very sophisticated as technologies go (an exception being the concentration stations and photovoltaic conversion), but their newness entails a number of drawbacks:

— unstabilized technologies subject to constant improvement make any transfer of technology difficult to accomplish;

— rapidly falling prices encourage a wait-and-see policy;

— technical and commercial competition between the solar technologies often shows confusion in the minds of decision-makers;

— the "optimized" technologies, which combine several renewable sources and sometimes conventional sources as well, are of very complex design.

HIGH HOPES

Along with these negative considerations, however, there is a rosier aspect:

a) in many of the developing countries, the solar and wind power resources and (except in desert regions) the vegetation are more abundant than elsewhere;

b) the cost of conventional electric power in the more isolated areas of the developing countries is excessive. I will not enlarge on the causes, about which you know more than I do; the cost of oil is only one of them, albeit an important one. Others are:

— the difficulties of transporting fossil fuels or the non-existence of an electricity supply network;

— the often modest size of the installations, generally about 100 kVA, which increases the cost both of investment and of maintenance;

— the irregularity of the load factor, linked with the cost of energy, which results in over-dimensioning of the installations in order to supply too short a peak load.

c) in view of the obstacles referred to above, the size of the solar stations, ranging from 10 to 100 kVA, is well adapted for the more under-privileged regions of the developing countries.

d) the very keen interest in solar energy on the part of the leaders of the developing countries; this is a positive factor that already holds out promise of collaboration between the industrial and the developing countries, and promises still more for the future. The developing countries can contribute knowledge of the real problems, and that knowledge is at least as important as the means of solving them. For we ought to be quite clear about this — there I am stating an opinion held by a person responsible for the solar programmes in an industrial country of only average sunshine, and it ought to be taken quite seriously by our guests from the Third World — I do not think that solar electricity can make a significant contribution to our own energy supply problems within a reasonable time — not, say, for the next twenty years — except in the one area that we share in common with the developing countries, that of biomass. On this subject I shall have a good deal to say today.

THE IMPORTANCE OF BIOMASS

Despite the fact that, at the onset of the industrial era in the 18th century coal and charcoal were the fuels that supplied industrial products, glass, metals and all domestic requirements, biomass now comes as something of a rediscovery for the industrial countries. It is otherwise in the developing countries, which are still extremely reliant on this resource; there biomass is as widely dispersed as solar energy, and usually regarded as a "non-commercial" source. In many regions of our planet, wood plays a vital role; it has the advantage of being crisis-proof, but demographic growth and the gradual improvement of living standards lead all too often to deforestation and desert conditions. It should be remembered, however, that wood is believed to cover 75% of requirements in tropical and equatorial Africa, 60% in south-east Asia and about 30% in South America. These data throw quite a different light on the statistics I gave you earlier in this report. Even in the industrialized countries, biomass accounts for 1% of consumption, which is enormous when we think of what it means in terms of absolute values.

I cannot lay too much stress on the importance of this vegetable resource, if used rationally and without excess; it is thought to amount to 200 000 million tonnes of carbon a year or ten times the annual world consumption. Without going into details, we all know of the good results that India has achieved with its digesters, or the drive launched by Brazil with its alcohol scheme. These are examples that provide food for thought for all of us, including those who come from the industrialized countries.

A FEW ECONOMIC CONSIDERATIONS

The generation of electricity by conversion of solar radiation is mainly done by thermodynamic or photovoltaic means. The case of biomass, a derived form of solar energy, will be considered later on. I shall confine myself to an examination of the existing technologies, leaving aside the concentration systems, for which the available data is inadequate for the purpose of forecasting any short or medium-term trends.

One often wonders which of the thermodynamic concepts or which of the photovoltaic techniques will win in the future. There is no denying that at the present time the installed low-temperature thermodynamic power in the developing countries is far greater than the photovoltaic power; since the economic balance today is around 5 kW installed power, the photovoltaic installations, numerous though they are, carry relatively little weight.

Moreover, the operating conditions of the two systems are not altogether comparable. In the case of the thermodynamic machines, any storage is in the form of hot water, which means that the maximum power of the installation is that of its a.c. generating set. With photovoltaic conversion, however, storage is in electrical form;

The output in continuous current is virtually unlimited, and in alternating current it is limited only by the converter. In the latter case, moreover, the losses in the battery (amounting to about 20%) do not occur when the battery is operating as a buffer, though they do have to be taken into account when operating in the charge/discharge mode.

By simplifying matters somewhat, however, one can make the financial comparison given in the table below. For this purpose the unit adopted is not the peak load, since this has hardly any practical significance: moreover, the peak power of the installation, especially in photovoltaic systems, is merely an engineering datum which, though important, does not reflect the real possibilities of power supply. The "nominal" power, dear to some electricians, has no significance unless the curve of daily demand is well known. A convenient concept for comparing these two types of installations is that of kWh supplied per day (understood as an average day). In the table, therefore, an attempt is made to extrapolate the investment cost in US dollars per kWh/day. There is assumed to be 24 hours of storage and very high-quality sunlight amounting to 3 400 hours a year.

TABLE 1

Comparison of 1979 and 1985 prices
(Investment costs in US dollars per kWh/day)
(with storage)

	1979	1985
1) Low-temperature thermodynamic (Station 50 kW peak)	$ 6 000	$ 3 000
2) Photovoltaic (Station 20 kW peak)	$ 7 500	$1 500

$/kW/day

It is assumed that:

1 kW peak → 6 kWh/day panels
5 kWh/day battery
4 kWh/day alternating current

In all probability the cost of photovoltaic energy will fall below that of thermodynamic energy in the years ahead. The relatively slight drop in prices envisaged for photovoltaic energy may come as a surprise, however, when we currently read in the literature of much more substantial drops of a factor of 10 at least.

Table 2 gives an approximate breakdown of prices for a medium-size photovoltaic installation (10 to 100 kW peak in 1979 and 1985). The values for 1979 are real values relating to a very isolated site at which the civil engineering, installation and transport costs are very high. In the 1985 hypothesis it is assumed that the basic infrastructure already exists in the area in question, that the fitters are trained on site and that a local management has been set up.

TABLE 2

Breakdown of prices for a photovoltaic installation
in 1979 and 1985 (US $)

	1979	1985
Cost of kW peak	(10 m²)	(8 m²)
Encapsulation	10 × 450 = 4 500	8 × 100 = 800
Silicon	8 × 400 = 3 200	6 × 60 = 360
Manufacture	8 × 300 = 2 400	6 × 30 = 180
General expenditure, profits	5 100	660
TOTAL PANELS	15 000	2 000
Batteries (24 hours)	1 000	800
Civil engineering structures	3 000	500
Convertor	2 000	300
Installation	1 500	300
Transport	2 000	1 000
Profits, sundries	5 500	1 100
	30 000	6 000
Cost of Kilowatt hour		
Interest 6%	Life: 20 years	(batteries 10 years)
Annual instalment	2 500	500
Maintenance/year	500	100
	3 000	600
COST OF kWh	2	0,40

This table calls for two comments:

— the prices given here allow the contractors adequate coverage for general expenditure and profit. This is a cost factor that is too often forgotten in the estimates found in the literature, which usually give the cost price only. The margin allowed may even be rather slender in the case of export operations to distant countries, where the risks can be extremely high;

— Although the prices for panels have been reduced by a factor of 7.5 (I cannot caution too strongly against over-optimistic forecasts), the total price drops only by a factor of 5 owing to the fact that in the non-photovoltaic part of the installation there are less easily reducible items.

Alongside these two examples of the direct use of solar energy, it is illuminating to set the investment costs for generators based on the gasification of biomass, a technology that is now well run in.

Table 3 gives the 1979 investment cost for a station of medium output — a few hundred kVA — still assuming it to be installed at a remote site. This investment is very low — 25 times less than for a photovoltaic station — but it should be pointed out that the kVAs of a gas engine have an energy value several times greater than the kW peak of a photovoltaic installation. All the same, we must add to these investment costs the cost of collecting the biomass at the rate of several tonnes a day, a cost factor that can be estimated only in relation to the special circumstances of each installation.

In Table 3 we have given a rather extreme estimate of what might be the cost, a few years from now, of an electric generator running on producer gas and using an engine derived from a car engine. It should be remembered that a car engine is the cheapest power generator available owing to the large scale of production. It costs something like 10 US dollars/kVA (provided that the alternator, too, is series produced).

An engine of this kind adapted for low-grade gas will produce less power, but its price will still be highly competitive. The only drawback to this solution, which still has to be put to the test of experience, would be its life-span. Large fixed engines have a service life much longer than that of car engines, which last barely more than 1 000 hours.

TABLE 3
Biomass gasification
100 kW Station (1979)

Investment per kVA	
Gas producer	300
Engine	800
Alternator	100
Installation	200
Transport, sundries	400
	1 800 (US dollars)

+ collection of the biomass (1 kWh → 1 kg dry biomass)

Forecast 1985 (??)

50 kW station	
Modified car engine	
Gasifier with charcoal	
Gasifier	20
Engine	30
Alternator	20
Installation, transport	30
	100 US $/kVA

Service life?

CONCLUSION

To conclude, I will refer back to a number of points which, in my view, ought to be taken as the guidelines for any cooperation in the generation of solar power; these remarks are addressed to the industrial countries and the developing countries alike:

— the funds allocated in the national budgets of the industrialized countries to the development of solar energy must be accompanied by a genuine wish to make these technologies accessible to those who most need them, namely the under-developed countries, at any rate for the next twenty years or so. Their technological effort should therefore include making training and information as widely accessible as possible;

— on the other hand, it is in the markets of the developing countries that information on decentralized energy requirements must be gathered. The experience of those countries will be invaluable to everyone. This means, however, that they will have to carry out continuous programmes of their own — programmes which will require less money than for technological development, but just as much intelligence if they are to make an effective contribution to the promotion of solar energy. For their experience to be useful to all, they must make it a rule to study the economic and social impact of their tests. It is only in the language of scientific experiment that they can usefully disseminate the knowledge they have acquired.

Returning to the industrial countries, let me stress that they will not necessarily be acting out of pure altruism in making their technology available to the underdeveloped countries. In some of the western countries, we are confronted with the gravest challenge we have had to face since we embarked on the type of civilization we know today — that of a foreseeable shortage of energy in the long term. Among the possible solutions (and I have no doubt there will be a whole range of solutions which we shall have to combine one with another), there is every likelihood that solar energy will have its place, ill-adapted though it is to our centralized systems of energy production.

These technologies will have to serve a long apprenticeship: initially their application will be found in satisfying the requirements of the developing countries, and it is in the best interests of all of us to join forces in order to achieve that goal.

B. SUMMARY OF THE SESSION

by

W. Bloss, General Rapporteur
J. Gretz, K.H. Krebs, J. Pheline, F. Jäger, Rapporteurs

Electrical and/or mechanical power derived from solar energy is an important factor in providing better living conditions in rural areas. There are applications in the residential sector, in agriculture and small-scale industry and in the field of transportation, as indicated in table 1. This table also includes the minimum power requirements of individual systems.

TABLE 1

Applications of Electricity in Rural Areas

Lighting	0.1 kW
T.V. Sets Communications	< 0.1 kW
Water pumping	1 kW
Cottage industry Small agro industry	5-10 kW
Deep freezing and Refrigeration	< 80 kW
"Solar Village"	100 kW
(500 people) Electric Power	20 kW (No Cooling)
Transportation	> 20 kW

All applications of power systems should be primarily based on the local requirements which can be met by different types of power generators. An analysis of the local situation has not only to take into account geographical and climatic conditions but also characteristic data of the special requirements for energy and specific load patterns. Although electricity represents only about 3% of the energy needed, its important contribution as high quality energy to an integrated system has to be properly estimated. Total energy concepts utilizing solar energy represent an ideal approach for further development.

The different methods of power production have not only led to a variety of different devices which are highly interesting from the scientific and technological point of view but they also offer a broad spectrum of power scale with different energy input which thereby promise an optimum matching to the local requirements. These methods have also been reviewed for the special application of water pumping where a power range in the order of 1 kW has been envisaged.

1 Thermodynamic power generation

Thermodynamic power generators utilizing solar energy have been used in a number of plants in the power range above 1 kW. In this type of generator, heat collected in absorbers from solar radiation is converted into mechanical or electric energy by turbine or piston engines. Flat-plate collectors provide operating temperatures of less than 100°C thereby limiting the efficiency of conversion to values below 5%. Low concentrating systems (e.g. strip mirror collectors, parabolic trough mirrors, Fresnel lenses) provide higher operating temperatures at about 300°C, whereas the use of mirrors which focus solar radiation to a central absorber enables temperatures of 500°C to be attained and conventional turbines to be used ("solar power concept"), as indicated in figure 1. It is generally agreed however that the use of concentrating devices requires frequent skilled maintenance.

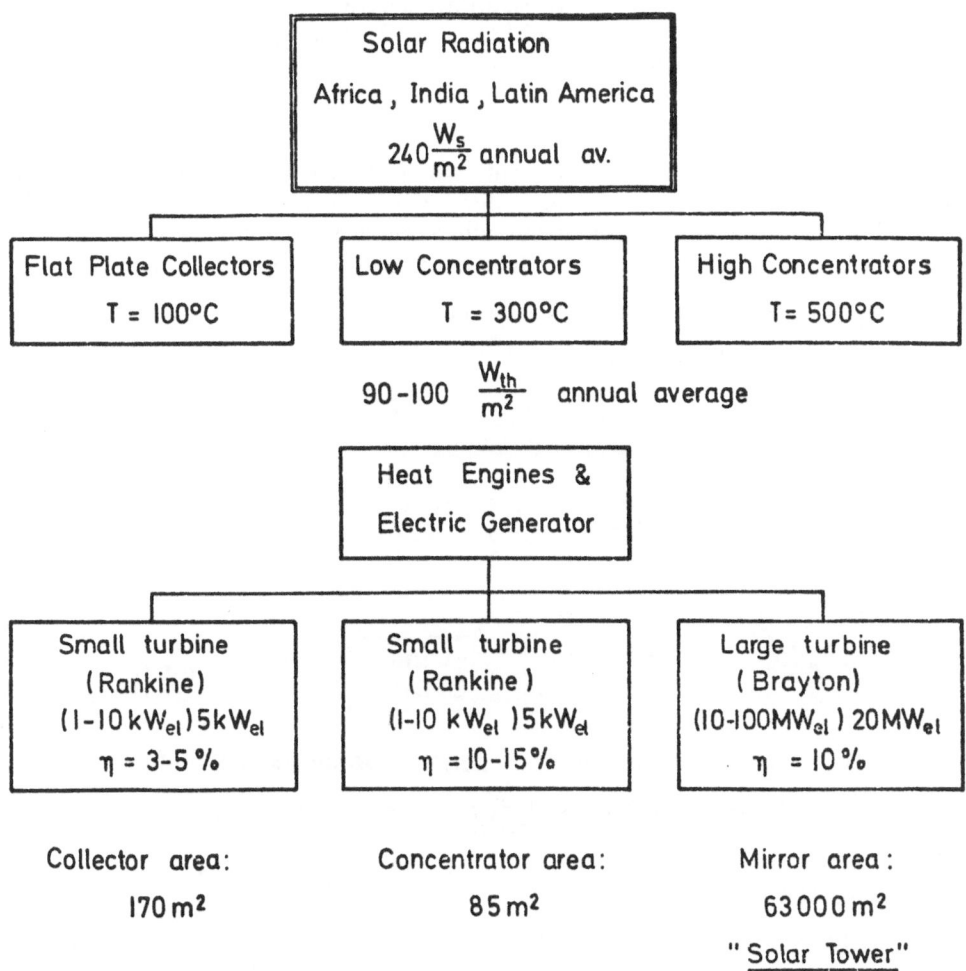

Figure 1: Thermodynamic Power Generation

Since these solar thermal plants use direct solar radiation, the problem of storage — thermal or electric — arises in consideration of a number of applications.

It has been pointed out that for solar thermal power generators even in mass production a lower limit of investment costs of about 8 000 $ kW_{el} could not be undercut if manufactured in industrialised countries. By local manufacturing, using appropriate technology, lower values of investment costs might be achieved. As an approach to reduce the mirror cost, a concrete shell/reinforced fiber technique has been developed in Indonesia.

It has been indicated that piston engines are better suited than turbines to tough field conditions. Generally the reliability and lifetime of solar thermal devices has to be increased and a large number of in-field tests, giving experience in operation and maintenance, are required.

2 Photovoltaic Power Generation

Solar radiation can be directly converted into electricity by solar cells, which consist of semiconducting crystals or thin films (fig. 2). Energy conversion is achieved by a junction between different semiconducting materials, in which separation of light-generated charges takes place. Research, development and fabrication have been focused on silicon solar cells.

The major drawback of high investment costs can be overcome by the following strategies: 1) development of low-cost silicon solar arrays; 2) use of thin-film cells; 3) development of concentrator arrays which replace high-cost cell area by low-cost optical concentrator area and reduce the system size by high efficiency.

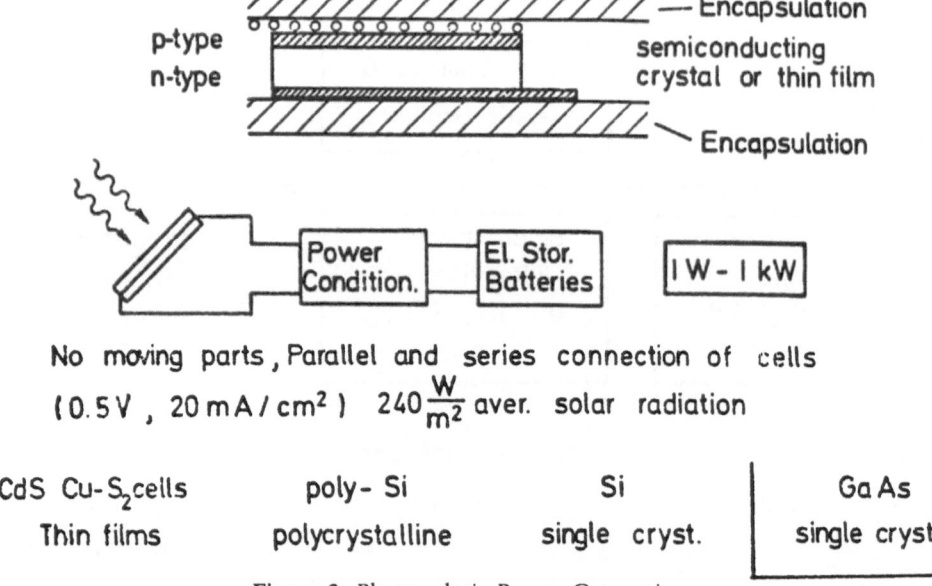

Figure 2: Photovoltaic Power Generation
(Solar Cells)

Since no moving parts are involved, a maintenance free, long-life operation can be expected. Special features of photovoltaic generators are their modularity, which provides gradual increase of the system. The intermittent or continuous power output depends on the capacity of the storage batteries.

Users

Country	Use	Remarks
India	water pump Si cell pilot line production	79/80 : 10 kW
Ivory Coast	educational TV	
Jordan	telephone line	132 units
Malaysia		
Mali	hospital (San) water pumps	8 kW 40/50/80 m³/day
Mexico	Si cell pilot line production water pump, TV	600 W
Niger	telephone grid educational TV	several 100's
Pakistan	lighting	1-2 kW (planned)
Zaire	telephone line	50 units at 100 W (planned)

A number ùf applications have been identified in 2 phases depending on the investment costs.

Phase I (costs $> 2\,000$ \$/kW$_{el}$)

Photovoltaic Power Generators in the range of 100 W:
 educational television
 telephone lines
 lighting of rural homes and public buildings
 equipment for science laboratories in rural secondary schools.

Photovoltaic Power Generators in the range of 1 kW:
 combined use of above mentioned applications, water pumps
 agrotechnical equipment.

Phase II (costs $< 2\,000$ \$/kW$_{el}$)

 water pumps
 cooling and refrigeration
 driving motors
 central power plants in villages
  (power range 20 to 100 kW).

It has been concluded that generally more in-field testing of photovoltaic systems is required to prove reliability under different climatic conditions, and a large number of small systems for different applications should be installed.

For the operation of larger photovoltaic generator systems a small number of projects has to be identified which fulfil the following criteria:

— selected application is of wide interest and will meet a real need
— probability of success is high
— complete and continuous long-term monitoring is possible
— information is available.

Pilot installations and demonstration projects serve several purposes. They bring cost analyses to a more realistic base, facilitate the assessment of technical credibility, reliability, performance and acceptance and bring to light failures and problems of installation and maintenance. Generally, photovoltaic power generation demands more extensive efforts in research and development to bring costs down to an economic level.

Concern has been expressed that the potential users in developing countries may by strongly dependent on manufacturers from industrialised countries.

At various places in developing countries, however, research and development work on solar cell arrays is going on and pilot production lines have been implemented, which demonstrate the feasibility of locally manufacturing photovoltaic devices.

3 Power Generation based on biomass

Power generation based on biomass deserves high interest since biomass represents one of the greatest energy potentials for tropical countries. The use of biomass for power production depends strongly on local conditions. It is derived either from special energy plantations, from agricultural residues or from industrial waste and can be used in stationary power generators and in vehicles.

One sixth of the world's fuel supplies are wood fuel and half of all the trees cut down are used for cooking and heating purposes. In the non-OPEC developing countries, which contain over 40% of the world's population, non-commercial fuel often comprises up to 90% of their total energy use. This non-commercial fuel includes wood, dung and agricultural waste and because of its nature is seldom thoroughly considered, e.g. (1) total wood fuel consumption is probably three times that usually shown in statistics and about half of the world's populations relies mainly on wood, (2) supply statistics of non-commercial energy can be out by factors of 10 or even 100 times. The use of biomass energy has to be studied in close connection with other kinds of use, like food or industrial use.

Biomass can be converted by means of microbiological conversion (fermentation), pyrolysis and gasification into liquid or gaseous fuels which can be easily stored. It is shown in figure 3 that electrical power production can be achieved by heat engines, combustion engines or, as a future aspect, by fuel cells.

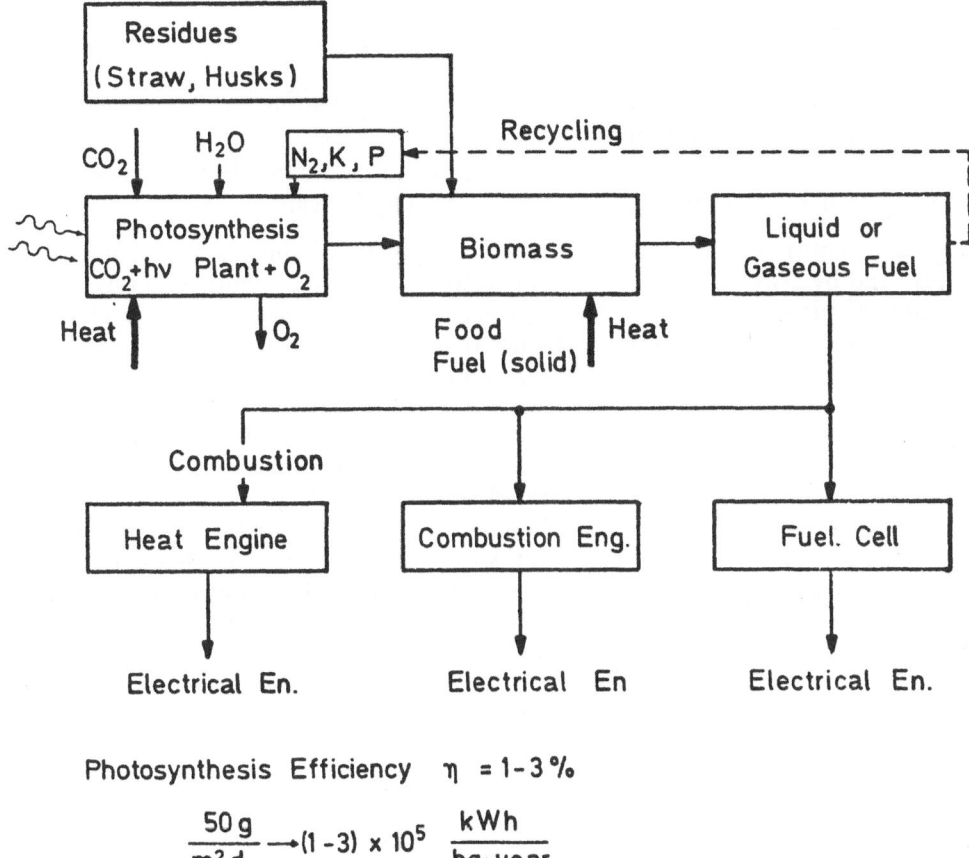

Figure 3: Power Generation Based on Biomass

In view of the diversity and complexity of the subject the following aspects have to be considered:

— local requirements for energy
— nature and form of energy products derived from biomass
— origin and scope of potential resources
— different processes of conversion.

It was requested that special attention be given to the methods of collection and preprocessing which frequently condition the total costs.

A number of examples were given to demonstrate the importance of biomass energy and to describe the most recent developments in this field:

— strategy of the Brazilian Government of large scale production of ethanol derived from different plants (sugar cane, manioc,...) and utilization as an alternative fuel in combustion engines

— methods of production and utilization of biogas for domestic applications in China

— Indian project on utilization of rice straw for combustion and power generation in turbines in view of industrial applications

— development of gasifiers and gas motors combined with electric generators in the power range of several 100 kW in different African countries.

A great interest was expressed in the development of the resources offered by forests and by selected cultures to improve the efficiency of photosynthesis and the location of plantations. The E.C. program of biomass research and development was described. It includes the energy use of straw and algae (Italy), central power stations (Ireland) and the study of economic aspects (assessment, cost, etc.). With regard to economic aspects, social-economic costs should be given priority over the pecuniary costs. It was concluded that the energy use of biomass could provide short term solutions to meet local requirements. Technologies for power production are available and have been applied regionally with good results. Further an increase of efficiency in the use of biomass is desirable and special attention has been focused on the improvement of stoves, combustors and digesters. It was requested that the number of demonstrations should be multiplied and special attention given to the use of agricultural waste. Production of liquid fuels in some cases exceeds the demand for application in rural areas. This excess can be considered as being available for alternative fuels to power combustion engines, which could have an impact on transportation problems.

4 Wind and small scale hydraulic power (fig. 4)

Wind power and water power both have a long history of successful applications for processes such as water pumping, grinding and mechanical sawing. In many countries the traditionally slow running wind mills for water pumping have played an important role in the agricultural development of remote areas. For example, in Argentina and Uruguay wind mills with a total capacity of 300 MW are estimated to be in operation today and in the United States of America more than 5 million wind driven water pumps have been installed. During the times of the easy availability of transportable fuels and the extension of electricity grids, many traditional devices have been abandoned all over the world and replaced by fuel consuming machinery. With energy growing scarce and increasingly expensive, it appears to be worthwhile to consider actively a revival of such techniques wherever the climatic and topographical conditions are favourable. It is felt in general that provided sufficient wind and water potential are available (coastal and mountainous areas), that wind and water power generation represent a very efficient and economic energy supply to rural areas. A cost comparison with the other solar energy conversion techniques under discussion during the conference has indicated wind and water power generation to be more economically attractive.

Wind generators for electricity generation have as yet received less attention as compared with mechanical systems and are currently under development in a few countries. Design, building and field testing of wind electricity generators has been initiated in countries like Chile, Argentina, Venezuela, Pakistan and Barbados.

Recently a variety of new designs making use of modern technology and materials have been studied.

The cost of systems is mainly governed by the construction of the tower. For small scale power production the costs are between 600 and 1 000 $/kW$_{el}$. Local manufacturing and appropriate technology play an important role in implementation and result in very favourable costs of 300 $/kW$_{el}$.

Modern small windmill designs, vertical axis and horizontal axis, are more economic for electric power generation (1 kW). But they need field testing to improve their reliability and adaptation to allow for local manufacture.

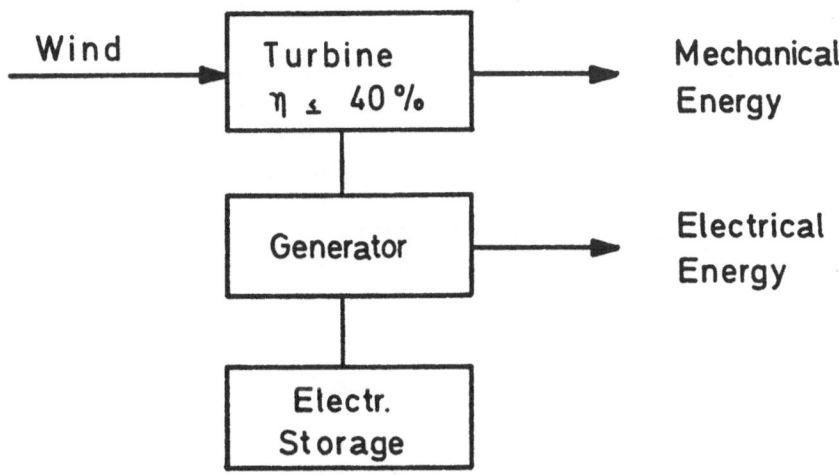

Cost estimate :   600 - 1000 $/kW   Wind power
1000 - 2000 $/kW   Water power

Figure 4: Wind Power or Small-Scale Hydroelectric Power Generation

It was noted that construction materials need further improvement with respect to durability. However, it was emphasized that locally available materials should have priority even with lower durability, since manufacture and repair of components would support local employment.

Discussions led to the following recommendations:

— Field testing of modern designs aiming at low costs should be started in order to obtain experience on system performance and reliability.

— In parallel with field testing, a systematic measurement of mean wind speeds and velocity duration distribution is recommended.

— The opportunities for local manufacture (including use of local material) should be investigated. Extremely favourable costs were quoted as being achieved by local engineering (300 $/kW$_{peak}$).

— Medium-scale wind generators in the range of 20-200 kW are considered to be promising to supply electricity to local industry or mini-grids in villages.

— Since quite often wind availability goes along with bad weather and lack of direct solar radiation it appears to be advantageous to complement systems converting direct radiation with wind generators.

Water power has been applied for many years to drive traditional types of machinery like mills or sawing plants. The generation of electricity requires higher speeds of rotation which can be achieved with small scale turbines like the Pelton or Banki type. Small scale usually refers to a capacity range of ten to a few hundred kW. This power level is appropriate to provide electricity on the village or small industry level respectively. Design and manufacture of such small scale water turbines can be well within the capability of local engineering. This has been demonstrated by experience reported from Pakistan, where local construction of two Banki turbines has led to remarkably low costs of 300 $/kW.

Although water turbines have to be designed to the locally prevailing hydraulic conditions to obtain optimal efficiency, standardization might be more important since the problem of spare parts would be eased.

In many cases a dam is required to augment the available head of water. Dam construction implies additional investments for capital and material and can possibly be avoided by extending the intake channel upstream or by applying new concepts like the zero head river turbine.

A number of recommendations have emerged:

— Engineering for local manufacture should be considered since experience indicates that due to high labour content in turbine manufacture, remarkable cost saving can be obtained.

— Hydro electricity generation is limited geographically. However, in sites where sufficient hydro potential is available it represents the most economic way of small scale electricity generation.

— The development of new concepts aiming at low costs such as zero head river turbines is considered as promising and deserves further examination.

CONCLUSIONS

Power production from solar energy can substantially contribute to improve the conditions of life in rural areas. Most of the devices are still in an experimental or developmental stage and combined efforts have to be made towards implementation. All types of power generators demand extensive field testing. Especially in those cases where small generators are technically feasible, a large number of devices should be installed in different climatic regions. The results and experience should be documented and efficient information exchange should be established. This information should not be only available on a scientific level but should necessarily include practical experience in operation and maintenance.

VII. Solar Heat

A. KEYNOTE SPEECH

by

Professor A.A.R. ELAGIB,
Chairman, Scientific & Technological Research Council (NCR)
President, Sudan Engineering Council
and
Member of the Executive Council of the World Energy Conference

Ladies and Gentlemen,

1. FOREWORD

On behalf of many of us who come from the developing countries I would like to thank the Commission of the European Communities for organising this forum for the discussion of a vital topic which is of high priority to developing nations. I am personally indebted to the organisers of the Conference for the confidence and the honour bestowed upon me by the invitation to address this learned gathering.

It was 18 years ago when I attended a similar International Conference on "New Sources of Energy" organised by the United Nations Economic Council and held in Rome. That was the first time I had a chance to meet some of the pioneers of Solar Energy work- the late Professor Daniel, Professor Baum, and many who are still hoisting the banner like Professor Trombe, Professor Perrot, Professor Delyannis, Professor Duffie, and George Löf. I pay tribute to the dead and send best wishes to the living. Judging by the achievements made by such early workers with the meager resources they had at their disposal, one is bound to have great expectations for the future, with the great resources now being made available for solar research, and above all the unanimous interest and support it has acquired.

It is gratifying to see now the participation of many nationals from developing countries. At the time of the Rome Conference there were very few and from a limited number of countries. Development can hardly be achieved or accelerated without a strong infrastructure based on a sound energy sector.

It can easily be seen that the level of development of any country — as represented by the GDP — has very close relation with its energy utilisation rate.

All the LDC's (less developed countries) invariably lie at the end of the curve. Note the effect of cheap hydroelectric power on significant raising of GDP in case of SWEDEN, NORWAY, etc. On the other hand all developing countries, including the LDC's, lie geographically within the tropics and hence they are potentially richer in solar energy. In order that they make effective contribution to the world provisions and sufficiency in food supplies, and to the saving of the non-renewable resources of energy, cooperation for the development of solar energy technology is of paramount importance. Use of high-intensive energy fuels of depletable origin, can then be restricted, in all countries, for technological activities which conditionally require such high-intensive energy fuels. Solar energy, in fact, being a natural type of energy, is very suited for the natural rural life and activities which most of the people in developing countries are leading. Its availability in situ avoids the characteristic and endemic shortcomings in infrastructural capabilities, transportation etc., which prevail in these countries and amongst such rural communities.

2. SOLAR RENAISSANCE

In Italy, one of the countries in which the European Renaissance originated, it is not out of context to borrow the word Renaissance to stand for the solar awareness movement which this conference should advocate. In spite of the logistic and seemingly convincing arguments presented about the need for development of solar technology for the rural communities in developing countries, there is little appreciation by planners for such development, because of the presently relatively high cost of solar equipment. Solar installations should not be evaluated on the basis of economic feasibility alone, but on the basis of the educational strategic long term value. In the history of Mankind many civilisations flourished in temperate or warmer zones. (At least the national expenditure will not include the heating bill. Moreover it should be easier to get rid of the heat by appropriate designs or to redirect it for another purpose than to generate it to fight the cold.)

So there is no reason that a revival or solar renaissance should not be achieved if sustained efforts are made to publicise and demonstrate the great potentialities of solar energy, and hence to create a wide-based interest amongst the scientists and the peoples of the world at large. Such a renaissance may well give birth to new Leonardos, Watts, Faradays, Edisons and Marconis.

Resources, however vast they are, will mean nothing unless resourcefulness and innovation are mobilised.

After all, the coal deposits which fuelled the industrial revolution in many of the presently advanced countries, would have been of far less value, had it not been for the inventors of the machines which utilised steam power. The rich oil fields would have been still hidden underground and oil products only used for Alaa Eldin lamps or for asphalt mortar as have been used since the days of the Kingdoms of Babylon and Mesopotamia. It is the simple approach of Otto, Diesel and Ford who gave oil its present worth. There is now an unfortunate concept that innovation is exclusive for top class scientists with sophisticated equipment. This may be true for making a hydrogen or neutron bomb or a complicated Teta or Tokamak apparatus. But for a down to earth energy like solar energy, simple down to earth approaches may be more useful. Furthermore, the developing countries may be still living the days of the Renaissance, or the Industrial Revolution, when there is still room for the simple thoughts of a newsboy like Edison or bottle-washer like Faraday.

3. SOCIAL AND TECHNOLOGICAL BARRIERS AND LIMITATIONS

Developing countries or communities may seem to provide a good virgin soil to seed new ideas or technologies. However it must be realised that such communities, on the other hand, are held between the powerful jaws of traditions on one side and the lack of technical experience and infrastructure on the other. So, ideas or designs which have to be introduced to such communities must come through a tactful approach, with clearly defined and potently benévolent objectives e.g.:

(i) To seed a change but not to force it.
(ii) To answer an actual need or to solve a definite problem.
(iii) To introduce equipment which is simple to operate and to maintain.

Solar energy application on the other hand has certain technical limitations. The fact that it is colossal as energy, but dilute and intermittent as power, is well conceived.

4. OPTIMUM PLANNING OF THE ENERGY SECTOR

For the optimum planning of the solar energy sector or of the energy sector as a whole, a macro and micro approach should be adopted. The macro approach should aim at devising special models for different climatological regions: e.g.:

The Jungle
The Savannah
The Desert
and The Coastal Regions.

Such approach was adopted in laying a long term strategy for developing the energy sector in the Sudan. In addition to the conventional sources, characteristic models are suggested for each region, based on non-conventional sources of energy. For the jungle the model is based on biomass. A communal development is suggested, based on a boiler, tractor and a tunnel dryer. Wood and vegetable waste can be either burned or fermented, or gasified into biogas. In the savannah where large agricultural schemes are found, a project is formulated for burning the one million-ton produce of cotton stalks for power generation. This amounts to an equivalent of 150 000 tons of oil. It is now being burned in the fields to avoid passing pests to the crop in the following year.

For the desert regions, water pumps driven directly by solar energy or by wind-power are suggested. For the coastal regions, where the winds are relatively high, windmills are most suitable. Solar desalination plants are proposed for both the coastal regions and the desert where the water is brackish.

The micro planning, on the other hand, should aim at selecting simple designs observing the various limitations cited before, and following a general principle of fully-integrated social and techno-economic approach.

5. SOLAR HEAT

Solar Heat represents an important part of solar radiation. Its real importance, however, lies in the fact that, practically all other radiations eventually, change into heat. I do not think that all possibilities in this area have been investigated yet.

A lot of heat is usually absorbed by the land and water masses, and radiated at different rates. In a desert during the night the temperature of the surface sands may drop to few degrees above zero while underground water and soil preserve a much higher temperature. Experiments in the University of Khartoum showed that if a slice of underground soil is well insulated from the sides, and the underneath, while the surface is covered with glass, temperatures of not less than 65 °C can be preserved through the night.

Such differential temperatures can be utilised in many ways. Already at Khartoum University, a solid absorption domestic size refrigerator is being run successfully, utilising the temperature differential in the day and night cycles. More work should be done on the thermo-electric effects (the Seebeck, the Peltier and the Thomson) and on new material technology.

Solar furnaces or kilns are attractive assets for production of building materials. A pertinent question is whether they may be useful to supply energy for the Bessemer steel process.

(a) The Simple Approach

Designs that can evolve from the village level:

(i) The passive use of Solar Energy:

The solar radiations are the cause of the heat and of the cold. So by cutting them off or letting them in, the climate indoors can be regulated. The raised roof mud houses of the north Sudan are typical examples of suitable habitat design against the heat. The idea can be incorporated in modern houses by having double roofing i.e. the main structural roof and a simple raised shading roof of plastic, canvas or local materials. The roof is usually responsible for 50% of the heating. The sides can also be protected by sun shades or vertical screens.

(ii) Another example is the well-known water cooling earthenware jar. The traditional design has not been improved since its invention at the end of the Stone Age. The design can be improved in many ways e.g. by increasing the evaporating surface (by adopting a finning design) or by increasing the head of water on the pores to mate the evaporating rate.

(iii) By simple technique food stuffs can be decontaminated from dangerous pathogens like cholera wibrio, and dysentry and typhoid bacilli.
The solar tabag (Arabic word for dish) is a form of solar still with no water but containing a load of foodstuffs. The glass or the polythene in this case acts as a one-way valve for the solar heat. Heat transfer by the atmospheric air is also cut off. The temperature rises above 60 °C in about 1/2 hour killing

such pathogens in less than an hour. The same technique was tried for disinfection of the cotton stalks in the project mentioned before. It gave very good results.

(B) Integrated Socio-economic Projects

"To answer a real need or to solve a definite problem."

(1) *Umsafari Solar Community Centre.*

Umsafari is one of three locations in Western Sudan, about 200 km west of the White Nile, where there are rich underground reservoirs of heavy brines. The chlorides are found mixed with sulphates and carbonates and hence provide very good salt lick for livestock. It has both nutritional and medical value. A traditional industry existed for centuries for extracting these salts by using fire wood. In so doing, however, the natives aggravated the desertification disaster which had already been looming over the area.

The solar project is now aimed at giving them an alternative technique for extraction of the salt by solar desalination. A feasibility study showed that the salt can be produced at half the present cost. Hence the return can easily cover the cost of all the auxiliary social work proposed.

The village will also be supplied with water from the solar stills. Due to the non-existence of potable water in the village, the villagers used to spend a good part of their income getting water from another locality 30 miles away. It is also intended to install windmills and solar powered storage batteries to provide lighting for a community centre for literacy campaign etc. The village will also be an experimental centre for field-testing of other solar facilities.

(2) *Hillat Hamad: BILHARZIA Fighting Project*

Hillat Hamad (or Hamad Borough) is a village in cotton Gezira Scheme between the Blue and the White Niles. It is a 2 million acres irrigation scheme, and one of the largest in the world. It was discovered that 80% of the population of the village at the age of 20 are infested with bilharzia — the water worm disease (Schistosomiasis).

It is contracted through the canal water. The thermodynamic solar pump now installed in the village pumps water from a borehole, hence avoiding the source of the disease. Its capital cost is high, but with its dual social and experimental function it has a highly improved cost/benefit ratio.

6. CLOSING WORDS

Finally, I would like once again, to thank the Commission of the European Communities for organising this forum and we look forward for their appreciated help to developing nations by embodying the spirit and findings of this Conference in the Lomé and in the UNCST Conference on Science and Technology which is to be held in Vienna this autumn.

B. SUMMARY OF THE SESSION

by

B. MEUNIER, General Rapporteur
T.C. STEEMERS, B. McNELIS, E. ARANOVITCH, L. CUOZZO, Rapporteurs

The working sessions bore on the following topics:
— the cooking of foodstuffs;
— refrigeration;
— air conditioning in the home;
— the use of solar energy in agriculture.

The following general observations can be made on these meetings:

— The satisfaction of energy needs associated with the preparation of food is regarded by most of the Third World representatives as a first essential. Hence national or international efforts should be directed in the main to the solution of this problem, which is of such vital importance to the populations concerned;

— For the most part, the technologies relevant to this area are already on hand (improved cookers, solar driers, passive air conditioning techniques, greenhouses for agriculture, solar water heaters, and more besides). The difficulty lies in organizing the interchange between the countries concerned of the great wealth of experience already accumulated. Here there is little need for original reserarch—the main requirement is adaptation to the real needs of rural life;

— Though these technologies are mostly already operational, they are not being readily disseminated, knowledge being confined as it is within the universities and research establishments. In this field, a considerable drive will have to be made to popularize them;

— To support this campaign, it will be necessary to step up the number of demonstrations in the field, with the local population in attendance. The local educational system must also be fully involved in the effort;

— Many solar projects have been implemented, but there is a serious shortage of feedback information. Objective systems for the retrospective evaluation of these projects would be of the greatest value to all concerned.

The three prime objectives to be adopted are the following:

1) To halt the catastrophic deforestation that is affecting a large number of countries, whilst at the same time assuring each inhabitant of a sufficient energy supply for cooking purposes;

2) To ensure maximum self-sufficiency in energy for the agricultural or agro-industrial development schemes, thus reducing their growing dependence on imported energy, which is destined to become more and more costly;

3) Lastly, to rediscover the advantages afforded by passive air conditioning techniques in the rural and especially the urban home.

*

The recommendations which might be adopted in the formulation of an initial proposal, which is felt by many to be essential, concern the **preparation of national schemes to satisfy energy needs associated with the cooking of food.**

These national schemes, which one night entitle "energy and domestic cooking schemes", would have to be set on foot with all urgency by the countries threatened by deforestation.

The objectives of the schemes might be:

— to sensitize public opinion to this vital question;
— to draw up a programme of coherent action dealing with the problem as a priority issue; the action taken would have to tackle both the demand for fuel (for instance, by extending the use of improved cookers) and its supply (optimum management of the remaining national forest areas, reafforestation, the promotion of other fuels obtained from vegetable waste, the introduction of biogas, etc.);
— to define the internal and external measures necessary for the implementation of these schemes.

The developed countries could render invaluable aid in support of these highly urgent projects (aid in the preparation of the schemes, in the training of demonstrators, in the financing of demonstration projects or retimbering operations, and so on).

These "energy and domestic cooking" schemes should, of course, take their place within the long-term national energy plans, which in many cases have not yet been drawn up.

To achieve the objectives outlined above, the following recommendations have been formulated by the participants:

a) **Cooking**

— The promotion of more efficient cookers needs to be followed up intensively. To assist the countries concerned, it might be useful to draw up a short manual describing the numerous variants available, and explaining their respective performances, advantages and drawbacks.
— Biogas cookers look promising, even though there are still many problems to be solved; their design and development should therefore be resolutely continued.
— The preparation of replacement fuels from vegetable waste is another likely approach, bearing in mind the abundance of vegetable waste still unused. The most promising techniques would seem to be the compression of waste into briquettes, conversion to vegetable charcoal and conversion to ethanol or methanol. It is recommended that support be given to the development of these techniques, which should be speedily introduced as soon as they are operational.
— The present solar cookers are ill-adapted to user needs. All the same, it seems that some research ought to be continued; the supply of energy for school canteens might, for instance, be a first objective: here the techniques

envisaged would be based on the use of high-temperature flat-plate collectors producing steam.
— National management of national forests and large-scale reafforestation with the active participation of the local population obviously call for priority treatment; every effort must be made to support projects launched in this field.

b) **Refrigeration**

The need for refrigeration has been acknowledged by the representatives of the developing countries as an important one. It concerns in particular the preservation of fish, fruit and vegetables, medicaments and vaccines.

A number of technological solutions are envisaged:
— compression refrigerators powered by motors supplied from photovoltaic cells;
— compression refrigerators driven by thermodynamic solar motors;
— absorption refrigerators supplied by solar collectors. The existing solar collectors are thought to be too dear; a big effort must therefore be made to lower the cost and give them a very high level of reliability.

Two applications might emerge fairly rapidly:
— small photovoltaic refrigerators for the preservation of medicaments and vaccines;
— modified absorption refrigerators that can use optionally either oil or gas (biogas).

c) **Air conditioning and heating in the home**
— The introduction of solar water heaters should be followed up more energetically. They should be brought into general use in public buildings such as:
  . hotels;
  . restaurants and canteens;
  . hospitals and dispensaries;
  . schools;
  . etc...
— Active systems of solar air conditioning are of interest but still too expensive. Nonetheless, their development is worth supporting, especially in view of the potential demand in urban areas.
— The passive air conditioning systems must be rediscovered and very largely re-introduced in the construction of dwellings and large buildings. The re-introduction of these techniques by architects would make for a substantial reduction of energy consumption in the large towns:
  . an inventory of passive air conditioning techniques should be drawn up on the basis of the broad experience already gathered in certain countries;
  . the architectural design of buildings calls for radical rethinking in order to avoid considerable wastage of energy and to increase comfort; the adoption of construction standards might also prove desirable;

further research should be conducted into passive air conditioning systems in order to arrive at a better appreciation of the relative utility of different options.

d) **Solar energy for agriculture**

The growth of agricultural productivity and the processing of agricultural products within the rural communities requires energy to be available at reasonable cost. In the long run, solar may turn out to be the best energy source to meet these priority needs. However, the development and assessment of solar energy should be carried out as far as possible on site, so that it can prove its suitability for local requirements and gain acceptance by the users.

These solar projects, whether isolated or embracing several technologies, must be integrated as far as possible with other rural development projects that hold out good prospects of success.

Solar energy, an omnipresent factor in agriculture, could quickly become operational not only in irrigation schemes (solar pumps), but in the following applications as well:

— *Drying of harvests and fish:*

   . Rapid and complete drying of the harvests does much to reduce product losses; the development and popularization of solar driers for the treatment of cereals, fish, tea, coffee, cocoa, etc., should therefore be maintained;

   . artificial driers supplied with heat from the combustion of straw and other vegetable waste would also be extremely useful for the rapid drying of products harvested in the rainy season.

— Controlled atmosphere techniques applied to the *heating or air conditioning of greenhouses* can also play a very important role in regions of difficult climate.

— *Refrigeration and preservation of products.*

— *Energy supply for agro-industrial undertakings:*

Agro-industrial undertakings such as dairies, canning factories or slaughterhouses are often sited in isolated rural areas; solar energy could go a long way to making them self-sufficient in energy, and projects on these lines would be well worth backing.

— *Soil sterilization:*

Soil sterilization techniques using solar heat are interesting in that they save using costly and sometimes toxic phytopharmacological products. Experiments in this field should be encouraged.

*

The participants have stressed the need to launch **demonstration programmes** in the field of solar energy as applied to agriculture. The demonstration projects, which should be carried out in a rural setting, must be given firm backing by the Governments and seek support from the various international organizations.

Once the technologies have been tested and found to be mature and compatible with local resources, **financial inducements** should be offered on lines similar to those that have been followed for the promotion of urban industries.

The participants have also expressed a wish that **very practical handbooks** should be prepared, setting out the state of the art in the following technologies:

. harvest drying;
. greenhouses;
. energy in the agri-foodstuffs industries;
. etc...

Report of the Specialized Technical Review Meetings

COOKING

Introduction

The importance of this subject is perhaps best indicated by the two following observations:

1° In several developing countries cooking accounts for a large portion of the total energy consumption.

2° Deforestation, as a consequence of firewood consumption, progresses sometimes more rapidly than nature can make up for. When this is the case in arid zones, the desert takes over.

Means

The means with which solar energy could help to solve the problems in this particular field are:

1° **Solar cookers:** which employ direct solar radiation either focused by mirrors or unfocused.

2° **Biogas cookers:** which use gas produced from cow dung, liquid sewage and domestical waste.

3° **Wood stoves:** improvement of the efficiency would reduce the need for firewood.

4° **Fuels from biomass:**

4.1. Production of charcoal: improvement of the carbonisation process.

4.2. Production of ethanol.

STATE OF THE ART

Solar cookers:

— several types of solar cookers have been developed and tested;

— some experiments to introduce these cookers in the rural areas have failed;

— traditional life style patterns form sometimes an obstacle for the introduction of this type of cookers.

Biogas cookers:

— tens of thousands of biogas plants have already been installed (and possibly millions in China);

— the biogas cooker has sociologically been accepted;
— the cost of a biogas plant is still too high;
— corrosion of steel containers is observed;
— the residue from the plant has a great value as fertilizer.

Wood stoves:

— efficiencies of 20% of the traditional stoves have been reported;
— Improvements with traditionnal techniques have experimentally proven to lead to efficiencies around 40%.

Fuels from biomass:

— the husks of coffee beans can, using a carbonisation process, be transformed into briquettes;
— alcohol can be produced from biomass.

Recommendations:

— development work on solar cookers should continue and field test should be encouraged.
Field test should however be preceeded by a study of the sociological acceptability of the solar cooker in the area concerned.
— the further development of the biogas cookers should be encouraged because of its great potential and because of the successful experiences so far.
In the further development and introduction of the biogas plant specific attention should be given to:
1° The reduction of the construction costs by using cheaper construction materials and cheaper construction designs.
2° The safety aspect; the installation should be fool proof.
3° Corrosion problems.
4° Training of construction and maintenance personnel.
5° Education of the potential users.
6° Financial incentives.
— development work to improve the efficiency of the traditional wood stoves should continue and the results should be made available to the people in the rural areas by means of an intensive demonstration programme.
— research and development concerning fuels for cooking and lighting based on biomass should be encouraged.

COOLING CHAMBERS, ICE PRODUCTION, REFRIGERATION

The application of solar refrigeration for the preservation of food and medicine was identified as of potential benefit in developing countries.

The possible technical approaches were discussed and the options for solar cooling are:

- Vapour compression refrigeration driven by photovoltaic cells
- Vapour compression refrigeration driven by a solar engine
- Absorption refrigeration driven by a solar thermal collector.

Additionally other related options which are considered to be interesting:

- Conventional or modified gas refrigerator powered by biogas
- Traditional passive water coolers (such as those described by Prof. Elagib).

The absorption cycle approach has received a lot of attention in the developed and developing countries and a number of projects were discussed. Of particular interest was the use of solid absorbent based systems one of which is operating at the University of Khartoum. An ammonia/water system for the production of ice has been developed in Papua-New Guinea. This produces 20 kg or ice per day and has an estimated capital cost of US $ 4 000 and produces ice at a cost of 10 c/PNG.

Refrigerators using photovoltaic cells are now commercially available. One has a capacity of 74 litres and is powered by a 60 W (peak) array and costs US $ 1 800. A similar 3 kg per day ice maker costs US $ 3 600. Systems which use the Peltier cooling effect are also under development.

The technical review meeting concluded that:

1. There are social problems associated with the use of village cooling chambers and ice production may be of more benefit because ice can be easily transported.
2. Systems should be highly reliable.
3. Cost reductions are necessary for solar refrigeration to be generally competitive.
4. The absorption cycle based system is the most attractive. In particular intermittent chillers offer benefit because of small requirements for auxiliary power. Development effort is needed in the field of collectors which must operate at around 100 °C.
5. Small Photovoltaic based systems are possibly the best approach for applications such as vaccine storage where ultimate reliability is essential.

The following recommendations were made:

1. Industrial development of solar refrigeration systems should be supported.
2. A series of demonstrations in developing countries should be undertaken. (For this suitable sites can be readily identified and the Food and Agriculture Organisation of the UN offered to provide data and assistance).

HEATING AND COOLING IN HABITAT

The subject comprises:

— production of domestic hot water;
— space heating;
— space cooling.

— in general terms, except for the production of hot water, interest is greater for urban areas than for rural areas, Solar Space heating is of secondary interest.

— production of domestic hot water can contribute:

- to improve health conditions even in rural areas;

- to save energy especially in large buildings, schools, hotels, hospitals, collective installations;

- to develop manufacturing; manufacturing of collectors is within the possibilities of developing countries, contributing to: utilization of man power; penetration of technology.

Lifetime of collectors is still uncertain (15 years ?).

Solar cooling

There is an increasing need for climatisation to:

— improve working conditions in new offices and working zones;

— to meet requirements of tourism and foreign visitors;

— up to now climatisation is covered essentially by electricity leading to peak loads which in some examples are already difficult to meet.

— but active solar cooling requires advanced solar technology i.e.:

- absorption machines;

- selective surfaces;

- eventually concentrating collectors with a tracking system.
  which is today uneconomic but interest nonetheless remains strong.

— the possibilities of evaporative cooling should not be neglected;

— to reduce the contribution of active solar cooling, passive systems should not be forgotten meaning returning sometimes to traditional architecture, better adapted to local climatic conditions than western imported techniques.

SOLAR ENERGY IN AGRICULTURE

I. INTRODUCTION

1) So far agriculture has suffered from lack of prestige and money and has not been considered as a basis for supporting the economy in many countries, it has to be recognized as the most important industry in most developing countries because food is still a most crucial problem.

2) If the productivity of agriculture is to be increased and most food processing to occur at the local level (a desirable local industry), **more energy** will be required for production and processing.

3) Solar energy is one of the most promising ways of providing such energy, but development and demonstration must be done at "village" level to determine which energy system will be more suitable to local needs.

4) Because of the seasonality of most agricultural operations and the very close association of agricultural operations with the village lifestyle and habits, **integrated systems** or **multipurpose solar equipment** should receive special consideration.

II. STATE OF THE ART

According to the present status of solar technology there are a few agricultural operations for which such technology might prove to be mature.

1) **Crop drying:** it seems to be the most common agricultural operation. At all latitudes and climates the timely moisture removal has many advantages over open ground drying whether the product be fruit, fish, coffee, tea or corn. Portable units or straw powered furnace operating dryers or biomass furnace dryers have been considered very appropriate.

2) **Controlled environment techniques,** both for heating or cooling greenhouses.

3) Cooling and product conservation

4) Water pumping, irrigation and desalination } topics covered in other session.

5) Intensive animal farming, for chicken, pigs, cows etc, where low enthalpy heat is required, compatible with the present solar technology.

6) Process heat, for agricultural product transformation "in situ" (dairy industry, wine distilleries etc) or other operation like soil sterilization, fertilizers production etc.

III. IMPLEMENTATION

1) To start introducing solar energy in agriculture a development and demonstration program has to be carried out at farm, village and cooperative level, matched with local skills and manufacturing capability.

2) **The demonstration program** has to be entirely backed by government, both national and regional, with the support of the various International Agencies.

3) Training and maintenance assistance should be provided, together with such a demonstration program.

4) After the technology has proved to be sufficiently mature and compatible with local resources, **financial incentives** should be arranged for the farmers and food processing industries to involve local entrepreneurs, in exactly the same way as financial support is given to industry in the cities.

IV. FUTURE POSSIBILITIES

We have gone through the exercise of outlining what is expected to become available in the near future, mainly tied up to the development of economical photovoltaic cell production or thermodynamic technique improvement. Some of the possibilities are:

— Battery driven small tractors;
— Solar refrigeration for food storing;
— Greenhouses with combined heating, cooling and water desalination;
— Air driven systems for combined heating & cooling operations;
— Solar driven crop sprayers.

V. RECOMMENDATIONS

1) A very practical and basic state-of-the-art survey and summary on crop drying should be made, assembling in a simple way all the information available on traditional and new solar techniques. The study should include simple construction drawings and diagrams to be passed to the village industries and farmers in a very comprehensive form.

2) Similar surveys and summaries would be worthwhile for greenhouses and process heat for the food industry.

3) A development and demonstration program, following a thorough technological assessment, should be started immediately, both for checking the technology at the field level and for training people to solve "solar problems". This would help to transfer the know-how from laboratories and universities to real life.

VIII. International and Regional Cooperation

A. KEYNOTE SPEECH

BY

F.A. DAGHESTANI
Deputy Director General
The Royal Scientific Society
Amman - Jordan

Mr. Chairman,

Distinguished participants,

It gives me a great pleasure and honor to have this opportunity to address such a distinguished gathering from various parts of the world. In dealing with the question of international and regional cooperation in the field of solar energy, one has to consider the overall energy situation on a global level. We are now undergoing an energy crisis that will continue into the future. This does not stem from the limitations of energy resources, which are abundent in nature, but from the very short time remaining to make these resources available. Lead times for energy systems development are long. Action is needed now to produce meaningful results by the end of this century. It takes ten years to start and operate a coal or uranium mine. It takes ten years to build a nuclear power plant. It has taken some 40 years to produce nuclear power that is now producing only 2% of world energy supply. The most recent breeder reactors, that could permit natural uranium to produce about 70 times more energy, will probably not be commercially available before 1990. Fusion technology is still at the stage of infancy and fusion reactors may not be commercially available in 50 years. All this lead time should be seriously considered in the light of increasing demand that is expected to increase 2.5 times before the year 2000 and to six times before the year 2050.

Energy consumption by developing countries has increased eight times since 1950 and their current share is about 16% of the total energy. The average per capita consumption in developed countries is about 13 times that of developing countries. This wide gap in energy share does not only exist between developed and developing countries, but also within each developing country. The major share of commercial energy is supplied to urban areas leaving almost nothing to rural areas inhabited by about 40% of world population.

Solar energy, as we have seen through the discussions during the last two days, holds great promise, particularly for the inhabitants of rural areas in developing countries.

The development and application of solar energy technology for the benefit of rural areas in developing countries requires serious and concerted effort at various levels.

Developed countries have recently intensified their research and development activities in solar energy with allocations estimated at 800 million dollars. Developing countries have barely started with meager resources and with an abundance of obstacles. Cooperation is needed at all levels to remove those obstacles and to build domestic infrastructure and know-how. Developing countries are markedly different from one to another and each country needs to achieve a capacity to make the technological choices relevent to its needs and to achieve self-reliance in an interdependent world.

If one wants to characterize the existing patterns of international cooperation between developed and developing countries in technical fields, one notes the following features:

1) Economic planners and decision makers in developing countries give low priority to cooperation in research and development. Technical cooperation is almost limited to equipment, studies by the foreign experts and some training.

2) Developed countries have only recently taken some limited action to strengthen the research and development activities in developing countries.

3) Variations of individual needs of developing countries were not given enough attention in technical assistance programs launched by the developed countries.

4) An integrated approach which takes into consideration social, economic and political factors in addition to technological ones has not been followed for the most part in many of the technical cooperation activities.

5) Foreign experts have given their best advise and in a sincere way. This resulted in adoption of economic and technological models unsuitable for local conditions.

6) Science and technology is recognized as an important element in social and economic development of developing countries. However, little cooperation has been accomplished in this field.

Having made these points, let me turn now to international cooperation needed in the future for solar energy, where a new and a fresh look is imperative. The features of such a cooperation in my opinion should take the following into account:

1) Cooperation between developed and developing countries should take the form of partnership where both parties should commit resources in the various elements needed to implement any project. Both parties should work together in the identification of needs in planning a project and to fulfil such needs in the execution of such projects.

2) Resource inputs from developed countries should focus on building and strengthening local capabilities to implement technical activities related to solar energy. Such a local capability should become institutionalized and the elements of continuity should be developed and maintained.

3) Individual needs of each developing country should be taken into account. Furthermore, local needs within a country should also receive due attention when planning and executing solar energy programs.

4) Developing countries should be encouraged to set up their own national science and technology plans in general and their solar energy plans in particular. This greatly helps in the commitment of local resources and identification of priorities.

5) More efforts should be exerted to identify those particular factors which act as constraints to the introduction, promotion, and adoption of solar energy applications in individual developing countries. Such identification of constraints should not exclude any possibilities whether political, social, or technological in nature.

6) Great emphasis should be made in planning such technical cooperation so that the end users of the results of such a cooperation will be the right sector of the inhabitants, particularly those described as the have-nots of commercial energy sources.

Having said all this, let me take the privilege of suggesting a vehicle for such a cooperation. Cooperation in the development of solar energy applications cannot and should not be considered as a marginal activity in the overall program of technical cooperation. This is why I would like to propose that the Commission of the European Communities establish an independent institution that may be named "The International Solar Energy Development Corporation".

This corporation would have a board of directors whose chairman is appointed by the Commission and whose members are equally selected from developing and developed countries. Those members would be selected in their individual capacity and according to their personal merit and knowledge of basic problems of rural areas in developing countries, particularly problems related to solar energy.

The board would be assisted by a relatively small permanent secretariat headed by a secretary general, whose task would be to execute policies and programmes approved by the board. The secretariat could also call for the services of ad hoc groups of experts as the need arises.

The corporation should have an annual budget of at least 50 million dollars to be provided by the Commission.

The activities covered by the corporation should include research, exchange of information, education and training and transfer of technology as related to solar energy for the benefit of rural areas.

Activities related to research would be:

1. To support research and development projects in developing countries;
2. To support joint research and development projects between institutions in the E.E.C. and developing countries;
3. To support joint research and development projects between developing countries;
4. To assist in the establishment of solar energy research units in developing countries;
5. To coordinate research activities through funding mechanisms;
6. To support research and development projects in E.E.C. countries when such projects are for the direct benefit of rural areas in developing countries;

7. To support research and development projects in regional solar energy research centers in developing countries when such centers are established;

8. To assist in developing solar energy research programs in developing countries;

9. To commission studies on the state of art of various solar energy technologies and to make them available to institutions in developing countries;

10. To support an exchange of researchers between E.E.C. countries and developing countries and between developing countries;

Activities in the area of exchange of information would be:

1. To make the results of supported research projects available to research institutions in developing countries;

2. To support regional and inter-regional seminars, workshops, and conferences with the participation of researchers from E.E.C. countries;

3. To assist in establishing linkages and exchange of information between research institutions in E.E.C. countries and institutions in developing countries;

4. To provide information to research institutions in E.E.C. countries on areas of research and development relevant to rural areas in developing countries;

5. To publish solar radiation data and atlases.

Activities related to education and training would be:

1. To provide scholarships for postgraduate studies or research in solar energy;

2. To provide short term on-the-job training scholarships in E.E.C. countries;

3. To provide short term experts from E.E.C. countries for training;

4. To provide training in research management, project planning, cost controls and reviews.

The Commission may consider an alternative to the organizational structure suggested here. However, this suggested corporation should be interested in all developing countries without restrictions to certain regions or sub-regions.

In conclusion, I would like to express my deep appreciation to the Commission for having undertaken all this effort to show concern for the needs of rural population. I hope that the Commission will not only submit a report to the United Nations Conference for Science and Technology, but also provide the vehicle for implementing what is in it.

B. SUMMARY OF THE SESSION

by

R. BATTI, General Rapporteur
J.K. PAGE, F. FITTIPALDI, A. REITHINGER, G. LIVI, Rapporteurs

All general activities pertaining to the implementation of the new solar energy technologies were dealt with during this session with the exception of social and environmental aspects discussed in Session 5.

The session was subdivided in four specialized technical meetings:

A. Radiation Data
B. Research, Exchange of Information
C. Industrialization, Transfer of Technology
D. Education and Training

The first speaker, Mr Maurice Foley, Deputy Director General for Development at the Commission of the European Communities, illustrated the Commission efforts to overcome development constraints.

This is being done through the Lomé Convention grouping 57 developing countries in cooperation with the European Community, through agreements with the Mediterranean countries, and an increasingly significant level of aid to the remaining developing countries. These development efforts focus mainly on food and energy. Large financial resources are available from Europe, including those to support regional schemes.

Although developing countries are chronically short of traditional energy sources, few requests for support from Europe in this field have yet been made. The developing countries, Mr Foley said, need practical application of new energy techniques. This practical application at the level of the rural social unit does not require new resources, they are already available. Therefore the existing opportunities should be seized and rationalized.

The new Convention presently being negotiated to replace the Lomé Convention, should provide a continuing opportunity for the developing countries to utilize European and their own experience to maximalize this important natural asset.

After the presentation of the keynote speech by Dr F.A. Daghestani and the introduction of the Working Document by the General Rapporteur a general debate was held.

Most delegates underlined the importance of regional and international cooperation for the future development of solar energy. Such cooperation should be planned on an equal basis between developed and developing countries.

Developing countries should not only become consumers of solar energy technology but should be involved at all stages of its development and use.

Several delegates have pointed out that cooperation should include joint design of projects and complementary production in the developed and developing countries by using endogenous capabilities and by developing indigenous skill in developing countries.

Solar energy equipment should as far as possible be manufactured on the spot, since in addition to providing employment, such equipment could be more easily set up, regularly maintained and repaired.

In view of the self reliance of the developing countries in the field of solar energy, endogenous industrial infrastructure should be strengthened through international and regional cooperation.

Many delegates have stressed the importance of access to regular and up date information. A great deal of scientific and technical information already exists, but there is a need to develop methods for its wide dissemination.

International channels are required for the transfer of information and many suggestions have been made during the debate.

Professeur A.R. Elagib suggested the creation of a regional organization which would be linked with the CEC and will serve to disseminate information on solar energy from developed to developing countries. The organization would have an advisory committee and publish a journal on solar energy which will serve as a mean to circulate information. The organization should also provide contacts between scientists from developed and developing countries.

Suggestions have been made to strengthen the contacts between scientists from developed and developing countries. Regular interdisciplinary conferences and seminars should be organized at regional and international levels to assess the work done on solar energy technology in the various countries.

Periodic news-letters on solar energy development should be published in regional languages such as Arabic, Indian, Spanish etc.

The dissemination of information from the research institute to village level has also been raised. Intensive national information and demonstration programmes would be a mean to provide the transfer of existing information and know-how.

Several delegates stressed the necessity to undertake socio-economic studies in order to assess the impact of new technologies such as solar energy on rural environment. Every new type of system must be introduced by taking into account all the social factors pertaining to a given environment.

The need to train specialists at all levels in solar energy has been raised several times. There is a particular need to train technicians to provide personnel for manufacturing, installation and maintenance. High level specialist courses, in-field training, and the introduction of suitable material in secondary and primary school curricula has also been suggested.

Regular information exchange of pratical experience should be arranged in regional seminars.

As pointed out by many delegates, education and training together with the strengthening of endogenous capabilities appear to be among the main factors needed towards developing countries self-reliance in the field of solar energy.

*

SPECIALIZED TECHNICAL REVIEW MEETINGS

Meeting A: Radiation Data

In view of the manifold applications of solar energy, a number of meteorological data are required which are beyond the usual frame of meteorological data registration.

In many countries equipment for radiation measurements exists, but this equipment covers few areas and is often inappropriate. Meteorological equipment for monitoring is too expensive, too sophisticated in maintenance and in data handling. Therefore it was suggested that regional stations well equipped and serving as local agents for collecting and distributing methodology and data should be set up.

Suggestions were also made to combine the data of the regional centres with satellite data in order to get more reliable information on the spatial distribution of the solar radiation.

Testing or development of cheap devices for meteorological monitoring with simple maintenance, and data acquisition and handling should be set up.

Some delegates suggested that close collaboration between the Commission of the European Communities (CEC) and the World Meteorological Organisation (WMO) should be encouraged. It has been suggested that a representative of the Conference on behalf of the CEC should participate to the WMO Congress scheduled for April 1979.

Meeting B: Research, Exchange of Information

There was a general consensus to recognize that, with a few exceptions, current efforts have not succeeded yet in projecting solar energy from the applied research field to the industrial stage. On the whole, solar energy is still an academic research activity.

It is important, therefore, that the research effort is continued and intensified to accelerate the process, as there is no doubt that, due to the inherent basic properties of this kind of energy, mankind should try to use it as much as possible.

In this respect, the contents of the working document is satisfactory. To proceed a step further, the following recommendations were made.

Regional research centres should be established among developing countries. The regional characteristics will suggest the appropriate organization pattern in each case, which will not necessarily be the same for all. The regional characteristics of the centres is a consequence of the fact that each region has its own social pattern and climatic reality. Also, there may be large differences between regions (and even between countries of the same region) as to the level of development and, consequently, requirements. Mr Mc Divitt (UNESCO) suggested to structure regional centres as a network of national ones doing related work rather than as a centralized institution.

Research in basic fields such as artificial photosynthetic systems of increased yield, or other new approaches, should be encouraged. It is important that research projects of this kind be located in developing countries when the local scientific situation allows it. Also, in joint projects between developed and developing countries, scientists of the latter should participate in order to provide a stimulus to local scientific activities of the country concerned in a project of medium or long range significance to this country. One delegate proposed that results from joint or individual work should be made available by publishing them in the literature.

Applied research should be continued, giving priority to the lines which appear more promising from meetings such as the present meeting or through panels of qualified experts convened for the purpose of analyzing the state of the art. The subject is relatively new and very wide, so it is important that technical meetings tend to detect the lines towards success in a reasonable time.

It is important that areas of pure and applied research are clearly differentiated from those of a development nature, for instance, subjects like forestation and biogas fall in the second category, though they may also lead to work in the first one. (Mr Harrison, UK).

As regards information, much of it does not reach CID or reaches it too late. In this respect, it was suggested that the EEC could help by publishing and distributing news-letters of brief collated information, in different languages, covering favourable and unfavourable results. The latter ones are very important to avoid unnecessary duplication in the case of applied work. The news-letters should give summaries information, with sources for further reference.

Assessments should be made of the prospective role of solar energy in the energetic economy of the country. It is important that these studies are framed in a more general study of energy policy in order to assess the real scenario and not be guided by hopeful expectations. (Ahmad, Papua New Guinea).

Examples of international collaboration were given and the necessity to handle them on an equal rights basis was stressed by several delegates.

Meeting C: Industrialization, Transfer of Technology

The subject gave rise to a large debate on several aspects of industrial and technological cooperation and in particular on solar energy. All the interventions have stressed the importance of strengthening endogenous technological capabilities in developing countries for this industrialization and for transfer of technology as well as for the development of the techniques, with a view

— to have access to technologies needed for this industrial national production and the needs of the export markets
— to adapt the techniques to local and specific conditions of each country.

In this context, it has been stressed that development and utilization of technologies are mainly a cultural and social problem.

This has been recognized as being particularly pertinent for the technological development in the field of solar energy. At the present stage of solar energy R&D

it seems unappropriate to apply conventional schemes for the transfer of technologies, i.e. transfer of technologies such as patents or ready-made equipment. A cooperation with participation of receiver countries is needed in

— research and development of the techniques
— the setting up of these techniques
— the adaptation to local and specific conditions of the receiver countries.

This would require to remove certain obstacles in industrial cooperation and transfer of technologies, mainly in the field of marketing policies, and also to increase bilaterally a more adapted assistance by international organizations and developed countries in favour of developing countries initiatives to create their own scientific and technical infrastructure at national and regional level.

Many delegates stressed the necessity to reach a New International Economic Order, which would solely be able to ensure that these problems be solved in favourable conditions for developing countries.

An invitation was therefore addressed to UNCSTD which will take place in Vienna in August 1979 to work towards a better adaptation of the structures of U.N. organizations with a view to realize these objectives. The CEC was invited to explore with its partners in the different cooperation agreements how these aspects of their relations could be improved.

Some delegates invited the CEC to take concrete initiatives to create a consultative organ for cooperation between scientists in the framework of the Lomé Convention or in a larger framework, or to establish a group of experts to coordinate projects, improve the information and give advices mainly in intermediate technologies.

In conclusion all participants stressed the importance of the following

— an increased participation of developing countries in industrial cooperation, transfer of technologies and technical development
— a better consideration of economic, social and cultural conditions of the receiver countries (developing countries)
— the necessity to evaluate each industrial and technological cooperation project for the development of endogenous capabilities of the receiver countries.

Meeting D: Education and Training

There was a general consensus to recognize that the availability of trained personnel, graduate engineers, research and technical staff are a prerequisite for carrying out research programmes and technical applications in the field of solar energy.

Many delegates stated that the local training in the developing countries should be a priority. Particular attention should be given to the training of national personnel in order to gradually replace technical assistance by national trained personnel. The need for local experts was pointed out for a better understanding of the socio-economic environment in the developing countries.

A number of delegates expressed the view that the efforts undertaken by several developing countries to reform their educational system should take into account the necessity to stress scientific and technical training.

Some delegates stated that a better way to realize research and training projects at the regional level should be done in connection with Universities and existing Institutes.

Delegates from developed and developing countries insisted on the importance of disseminating systematically the possibilities of training in the solar energy field and of establishing an exchange of expertise; it was suggested that this could be implemented via regional seminars or flexible liaison structure.

In responding to the debate the rapporteur pointed out that the Commission has concluded several agreements with developing countries or group of countries tending to reform their education and training system which could be considered as a reference for initiatives in the field of solar energy training.

He observed that the Lomé Convention foresees the cooperation between the Commission of the European Communites (CEC) and 57 ACP countries including also projects and programmes in the field of education and technical training. Each ACP country should define its priorities and objectives in this field.

Furthermore the agreements concluded by the Committee with the Maghreb countries (Algeria, Marocco and Tunisia) & the Machrak countries (Egypt, Jordan, Lebanon, Syria) provide possibilities of cooperation in education and training. Some solar energy research and training programmes have already been established and will be implemented in the near future.

Concerning the cooperation between the Commission of the European Communities and the non associated developing countries particularly in Asia and Latin America cooperation projects in the field of training could be envisaged.

He also stated that the CEC is ready to examine the possibilities of cooperation for research and training at the regional level, even in the framework of a horizontal cooperation (between developing countries) particularly in specific training programmes. In this context a cooperation between the Universities and CEC research centres and corresponding organizations in the developing countries should be strengthened. Joint Research Centre (JRC) would represent one of the main instruments towards the implementation of these projects.

IX. Environmental and Social Implications

A. KEYNOTE SPEECH

by

Prof. Djibril FALL
Institut de Physique Météorologique
Dakar, Senegal

1. PRESENT ENERGY SITUATION

For a few years now, we have been watching the increase of oil prices; this does not seem however to have substantially slowed down growth rates in the industrialized countries. An uncontrolled exploitation of fossile-energy sources will not only lead invevitably to their imminent exhaustion, but is also imposing considerable nuisances on the environment of man.

These facts will again confirm and stress that it is indispensable for both developed countries and developing countries to reorient towards new sources of energy.

Following the estimates published by the UN in Vol. 1 No. 6 of the "National Resources and Energy" magazine (dated June 77) of energy resources presently exploitable on an economic level (i.e. an equivalent of about 1.2 billion tons of coal, considering also the resources provided by nuclear energy), it is understandable that the majority of Third-World countries (representing 75% of world population in 2000) are feeling anxious about their economic and social development, which is largely dependent on energy supplies at acceptable conditions from abroad. In addition, poor ecological conditions and rather lamentable economies will also affect such anxieties.

All these factors will explain why it is hoped to exploit renewable energies, for example solar energy and its derivatives (wind energy, biomass, etc.) in developing countries, especially those with sunny climates and poor sources of combustion material. Even countries with abundant energy sources but little capital and few workmen to exploit such sources share the aforementioned hopes.

They were, however, delayed in materializing, due in particular to the slow progress of technology in this field. Until a few years ago, this was due mainly to the very little interest the industrialized countries paid to this problem for as long as there was lack of energy and prices for the latter were cheap.

However, the developing countries should not expect too much from the external world, just contenting themselves to merely use imported technologies. They should try to promote techniques of more interest and use to their specific needs with special attention to their resources and specific local living conditions.

Also any technology transfer should be subject to such imperatives and should take account of the possibility of making available new employment which is capable of calling for workmen's participation of the countries involved in the promotion of every technological project. To do so, the foremost task to be accomplished is therefore to set up appropriate training structures, research facilities and other instruments to achieve an efficient assimilation of such techniques. In connection with scientific and technical development in general, attention has to be paid at the same time to the specific political and economic, social and cultural characteristics of each individual country.

Substantial progress has been achieved during the last ten years and the number of users (active and potential) of solar energy is increasing from year to year.

If people are persuaded that resources of renewable energy will not alone solve every problem in the energy field, we are aware nowadays that a reasonable use of the aforementioned type of energy will help to satisfy a considerable part of the demand.

The main worries of our countries as to the problem of development are essentially concerned with the rural districts for the next few years to come. Rural districts represent the lion's share of our population. We have actually to concentrate our efforts on them as far as renewable energy uses are concerned, since here we find the areas of large extensions in surface with poor population density.

Many of these populations live in small villages, whose remoteness, as compared with urban centres, make fuel transport and power supply from the classical distribution centres extremely difficult. In this respect, we have to ponder over some data provided by the World Bank. In 1971, only 4% of the rural population of Africa, for example, were supplied power. By 1985, this percentage will still be below 10%.

It has to be said that the above data did not take account the increase of the prices of fossil-combustion material, which produce up to 45% of power in Africa.

The energy demand of rural populations considered separately is about some tens or hundreds of kW of installed power.

The solar energy need in particular and renewable energy in general will play an important role in this area if the implemented equipment is of a sufficiently robust nature, easy to handle and cheap and easy to maintain by semi-skilled workmen.

The increase of the energy supply to villages will represent an important factor in improving living conditions. Rural populations usually spend an average time of 2 to 3 hours per day collecting wood, cooking food, dipping water and grinding cereals. If these activities could be mechanized, more time would be gained to cultivate food products, improve the use of the land and cattle-breeding, to obtain more leisure-time, education and literacy, etc.

A rational use of solar energy is needed, along with an appropriate and patient education which should be gradual enough.

To abolish local habits, not too suddenly, might have a substantial effect on the struggle against the clearing of the woods and a better protection of the environment.

With renewable energies abundantly available, we are faced with a challenge and the effort to be made is within reach of our forces; the priorities have to be well established and objectives clearly defined.

External support, though badly needed nowadays, should not represent the one and only means to rely on. It can be of benefit for us in the long run only if we build the structures that will enable us to become independent as far as essential needs are concerned.

2. THE USE OF SOLAR ENERGY

The results obtained in research and application, as far as the use of solar energy and its derivatives is concerned, are promising and the activities under way should be more strongly supported by means of human and material help. Some types of solar equipment as implemented at present, have developed beyond laboratory level and are being publicly accepted. Solar water-boilers, for example, will be entering the commercial phase in urban centres and rural district structures (dispensaries, schools, etc.).

— Some tens of solar motors have been set up on rather poor lands, in order to pump up underground water within municipal and pastoral irrigation projects.

At the present stage, some of the aforementioned solar motors are placed alongside diesel motors of the same power in order to examine the problems of performance and their cost.

The results obtained are encouraging and offering prospects for the future especially as far as remote regions are concerned.

— The preserving of food products through solar drying is a vital question for our populations. Let me recall the fish that lose nutritive value under preserving. This example is quite edifying, as well as those agricultural products that remain the elementary food resource of our rural districts.

The improvement of drying conditions has been achieved in a large number of laboratories and the present phase is that of spreading the equipment developed in this way.

— In regions with a windly climate, wind energy can in many cases complement solar energy.

Research in this field is being mainly oriented along the setting up of practical and also relatively cheap equipment.

Solar television is about to contribute to education in rural districts, with programmes aimed at increasing literacy and improving social hygiene.

— Solar cookers are beginning to be introduced in some of the developing countries. Despite the curiosity it provoked, this device has not yet been accepted in a too encouraging way. Some local habits are hard to overcome, especially as far as cooking arts are concerned, and a very patient investigation will have to be conducted in the use of using solar or biogas cookers in rural districts, especially those where the energy needed is that required for cooking purposes. In Senegal for example, the State Secretariat for the Promotion of Human Affairs has undertaken some research programmes as well as an education action directed towards the rural population. The future will require initiatives of this kind, especially taking into account the problems of the desertification of the

105

Sahel region which are aggravated from year to year. Simultaneously with the use of solar cookers studies are under way as to how to use dead wood efficiently and how to reduce wood-cutting, considering our resources that have become poor in this field.

This action in the field of education will also have to take into account the caste societies prevailing in many regions of the developing countries. It will happen very often that craftsmen like blacksmiths, jewellers and shoemakers for example are considered to be at the bottom of the social scale despite all efforts made by public authorities to abolish such concepts.

Very few of the cultivators for example are prepared to repair their hoes, rakes or ploughs, even if they are able to do so. They call for the village blacksmith. This is a serious handicap to the promotion of appropriate local technology, consistent with the meagre resources of the populations concerned.

The expected results and their applications will, subsequent to the spreading of solar and other equipment in rural districts depend above all on the ability of those populations and their local craftsmen to produce and operate easy-to-handle, strong and cheap equipment.

In the field of solar energy, we are thinking above all of the concept of a very simple cooker, of a very small and elementary wind energy device as well as of ovens that could be manufactured by local craftsmen, so that the accumulated dead wood can be burnt more efficiently than has been the case up to the present.

In the medium range, the interest of these populations will have to be attracted to such activities that will enable them to play a more active role in the development of rural districts with a view to improving their conditions of life and in order to prevent them from moving to the cities where they hope to lead a more comfortable life than in the country. This will lead to overpopulating the cities and destroy any effort already made.

— The possibilities offered in the energy discussion by bio-gasification have led to the invention of some easy techniques that are well adapted to rural conditions and are already beginning to spread.

As a matter of fact, the industrialized countries realized the importance of the potential market offered by the developing countries for techniques they undertook to develop in the use of renewable energies. They are even prepared to set up plants to manufacture such equipment in those countries where the work is cheaper. All this should not make us however, remain inactive.

We need technologies, but they should meet our requirements, take account of our means and enable us to use our local material in the manufacture of the required equipment. Since it is not feasible at present to maintain sufficient competition with industrialized countries in the development of techniques (e.g. in the field of photovoltaics) for the efficient exploitation of solar energy, the developing countries have to be in a position to adapt imported technologies to their local conditions and to use standards which are appropriate in their view. It is in this connection that we have to face the vital problem of how to establish strong teams of researchers and technicians in well-equipped buildings.

3. FUTURE PROSPECTS AND CONCLUSION

With research and application work undertaken so far, with the help of mostly modest funds, future prospects are very promising, especially due to some countries' joining one another within subregions, pooling also their means and funds, thus facilitating research work which otherwise it would be very difficult to conduct by each country individually.

Within the means and funds to be implemented, the information sector is playing a preponderant role in view of the hardly feasible communication among developing countries. The latter are facing mostly the same problems, however, it is much easier for them to communicate with developed countries and to be informed on models as suggested by the latter. Plenty of countries failed with technical transfer upon such models. If only their neighbour states could draw some useful lessons from such failures in order not to commit the same mistakes and errors, thus having more chances to draw benefit from more interesting transfer.

The CEAO example is demonstrating and suggesting such a basis of collaboration between countries facing the same type of difficulty in their process of economic and social development.

Solar energy in particular and renewable energies more generally speaking will have an important role to play in the future; we have merely to avoid exaggerating what could be our optimism, since, though available free of charge, solar energy is not yet, and will not be in the near future, cheap enough in its exploitation.

It is therefore important to take good note of, and to thoroughly specify the ways and means to develop solar energy as well as the technical results aimed at in order to solve the problems of priority with reference to the prevailing energy crisis, the end of which cannot be foreseen at the present moment.

This planning work will have to select the most appropriate energy source; we know for example today, that solar biosynthesis, exploited scientifically, is one means offering great opportunities to overcome world famine.

The idea of using solar energy is no longer utopian nowadays, and the developing countries are not standing alone in this new challenge with the presently unfavourable energy situation. In addition to their vertical cooperation with industrialized countries they are now establishing a horizontal one among themselves, and they are being offered by international organizations like WMO, UNIDO, UNDP, UNEP, UNESCO, FAO, WHO, and others, an appreciable support in the way of material and experts.

The new challenge is first of all taking place as far as the encouragement of remote rural districts is concerned, and it is being proved that renewable energies that are completely adapted to a small-scale energy decentralization system might provide support to a more general programme of energy development.

B. SUMMARY OF THE SESSION

by

G. BEGHI, F.C. TREBLE

INTRODUCTION OF THE WORKING DOCUMENT

Summarising the content of Session 5 of the Working Document, Dr. Beghi pointed out that solar energy could contribute to the solution of environmental and social problems in the following ways:

- Reducing rates of deforestation and desertification.

- Arresting the migration of population from rural areas.

- Improving food production and agriculture.

- Improving education and communications.

- Providing on-site power.

- Preserving clean air.

- Avoiding harmful waste products.

However, certain conditions must be met if new or improved solar equipment is to be succesfully introduced without social and cultural disruption:

1. The equipment must satisfy a real need.

2. Possible alternative appoaches must be fully assessed beforehand, with regard to their efficacy, cost, reliability and ease of maintenance, their use of local materials and methods of construction and their adaptability to local customs and traditional habits.

3. Selected solutions must first be tried out in field tests and in situ demonstration projects, in which the users participate at every stage from the initial proposal, through the design, manufacture (where appropriate), installation and monitoring stages to final optimisation.

Demonstrations carried out on these lines will help to ensure the environmental, social and cultural acceptability of new technology, expose any faults or shortcomings in the equipment, familiarise the users with its operation and maintenance and serve as centres for further training. The results of such demonstrations, even when negative, should be fully reported and the reports distributed to all who could benefit.

Governments can do much to encourage the introduction of proven solar technology on a wide scale by subsidies, tax incentives, the removal of tax disincentives, demonstrations on public buildings and other techniques of financial assistance and market stimulation.

108

DISCUSSION

Opening the discussion, Prof. Hatry, President of the Energy Session of the Commission's Economic and Social Committee, stressed the importance of three requirements for the developing countries:

1. An education and training programme to develop professional skills and improve the quality of life.
2. An interest in developing manufacturing firms and training managers.
3. The need to maintain a constructive, open and sincere dialogue between countries, particularly in cases where it has become difficult to reconcile differing viewpoints.

A member of the ACP General Secretariat from Upper Volta supported the call for demonstration projects, with technical and financial assistance. Such entreprises, he said, should be placed under regional control. Information exchange between developing countries and between existing organisations was extremely important.

Regarding the choice of solar technologies, the solar cooker (despite some adverse experience to date) was advocated as a most important element in the prevention of deforestation and desertification, although it was pointed out that desertification could not be separated from the water problem. The hope was expressed that the growth of solar energy utilisation in developing countries would be helped not only by local manufacture of equipment but also by decentralised production of glass, insulating materials and other essentials.

It was recognised that, in some cases, the introduction of certain solar technologies would inevitably bring about changes in living conditions and habits and thus cause some social disruption. One view, from France, was that a full analysis of social and cultural side-effects should always be made before choosing a particular approach, although technologists could not be expected to find answers to all the problems. An opposing view from Bangladesh was that solar technology must always be applied in the most fruitful and effective way. Only after it had been in operation for some time would sociologists be able to analyse the consequential effects on living habits and propose remedies for such problems as would arise. It was necessary to separate the technological and sociological problems.

There was some feeling from Mali that, given good will, honesty and freedom from exploitation, the farmers themselves would make the right choice of technology on the basis of efficacy, cost and reliability.

X. Appendices

A. SUMMARY OF THE PREPARATORY SEMINARS

by

W. PALZ
General Organiser, Commission of the European Communities

In September and October 1978, 5 regional seminars were held with the purpose to prepare this Conference. The list of the seminars appears on Table 1.

Naïrobi, for East Africa

Bamako, for West Africa

Amman, for Arab Countries

Caracas, for Latin America

Delhi, for South Asia

TABLE 1

The topics of the seminars are summarized on Table 2.

Solar Energy Utilisation in developing countries, Emphasis on rural areas

- Status and outlook: practice and technology for solar energy utilisation.
- Current and future needs which could be met by solar energy.
- Technical and non-technical barriers to overcome.

TABLE 2

Each seminar lasted one week and was attended by about 20 experts from 7 to 10 countries. A report was produced for each seminar and subsequently a working document for the Conference which attempts to draw general conclusions. As this document has been widely disseminated already, this paper will restrict itself, to a short summary of these conclusions. A list of general results appears on Table 3.

111

1. In many villages, solar energy is still the only resource. Traditional ways of utilisation are very inefficient and endanger the environment;
2. Improved traditional and new technological approaches have the potential to meet the basic energy requirements of most villages. They are often better suited than fossil energies for rural development;
3. Needs are tremendous and pressing.
   They call for large and effective development programmes;
4. Not all technologies offer the same promise. Experts do not always agree in their appreciation;
5. Major obstacles are cost of products and lack of general infrastructure for development and operation of systems;
6. International cooperation is desirable.

TABLE 3:

General conclusions

From the seminars was concluded that the concepts and technologies which offer particular promise for future implementation are the following

1. Improved uses of biomass:
   — silviculture of selected species for arid lands;
   — larger use of available residues;
   — improved wood stoves for cooking;
   — development of biogas;
   — improved conversion methods (charcoal, ethanol ...)
2. Plastic greenhouses are important for agriculture.
2. Solar distillation is not yet well developed; solar stills are appropriate for small needs.
4. Photovoltaics hold out considerable promise for power production and water pumping.
5. Wind generators raise maintenance problems but are particularly suitable for local production.
6. Production of solar heating devices is well advanced in many countries.
7. Active solar cooling is very expensive.

TABLE 4:

Some technical issues

Experts did not always appreciate in the same way prospects of the different technical options. Some open issues are put together in Table 5.

- Has the "solar cooker" a future?
- Is biogas eventually the answer for cooking in rural areas?
- How practical are thermal power generators and water pumps with a view to maintenance needs?
- How practical is solar refrigeration for food conservation, given that acceptance problems of the villagers exist even for conventional cooling chambers?
- What are the advantages of "solar drying" in comparison with conventional air drying?
- **What is the priority ranking of solar water heating in rural areas?**
- Is it acceptable, given its scarcity, to employ water in passive cooling systems and in thermal power stations?
- Is series production appropriate if local transport is a problem?

TABLE 5:

Some open issues

There was general agreement that a number of measures should be taken urgently with an eye to the widespread utilisation of solar energy in the developing countries, and in particular in the villages. International cooperation is most important for the success of future programmes. Table 6 shows the tasks which deserve particular interest:

1. Solar radiation data:
   - conform to WMO regulations;
   - extend networks;
   - exploit satellite data;
   - develop simple devices for engineers.
2. **Assessment studies**
3. **In field testing to demonstrate the viability in local conditions**
   - cookers;
   - solar cells for lighting, telecommunication, TV etc.;
   - solar power plants in the 20 to 100 kW range;
   - solar water pumps with emphasis in the 1 kW range;
   - crop dryers;
   - small distillation units;
   - heaters for agro-industry;
   - integrated systems, e.g. the solar village.

4. **Scientific and technological development**

   Solar energy research adapted to the local competence and infrastructure and solar energy as an impetus to develop local competence and infrastructure:
   — Practical and cheap biogas units;
   — Solar collectors from local materials;
   — Solar dryers;
   — Passive cooling systems and others.

5. **Development of appropriate production capabilities:**
   — Centralised;
   — Craftmenship in the village.

6. **Institutional development**

   Possibly national solar energy centres linked to a regional centre.

   Objectif:

   Promote solar energy nationally vis-à-vis of the governments, in universities, research centres and industry and towards the potential user; international cooperation.

   Activities:

   — Dissemination of up-to-date information;
   — Testing and quality control;
   — Workshops and courses etc.;
   — Demonstration programmes.

TABLE 6:

Some general tasks which were identified

On the basis of these conclusions, it is hoped that this Conference will lead to a more comprehensive and detailed outline of activities and actions.

B. ENERGY FOR RURAL DEVELOPMENT

A Film depicting rural life and energy needs

by

Dr. H.N. SHARAN, Director
Bharat Heavy Electricals Ltd, India

The film shows scenes of rural life in India, the conventional methods to meet their energy needs and the changing scene. The text below gives a brief description of the theme on which the film has been based:

In order to concentrate our entire attention to energy needs for development, one must clearly understand the life pattern and life style in rural areas of developing countries. The rural population has developed over thousands of years, techniques and methods to fulfil their energy demands, methods which in many cases are appropriate to their needs and cost effective for their life style. Other traditional methods create serious problems and need improvement.

Energy for water production, for example, is one of the essential rural needs. In absence of modern methods and conventional power, villagers have depended upon human and animal muscle power to fulfil their needs for drinking water and irrigation from deep wells and shallow ponds. These techniques, based on various laws of mechanics, have been in use since time immemorial and it is interesting to note that these are being used even today all over the world.

Energy for cooking is another vital need of villagers. The most widely used energy source is firewood. The burning of firewood with conventional methods is not only inefficient but it also causes pollution and is highly labour intensive, needing 4-5 hours of one or two family members for its collection every day.

Where does the firewood come from? The rural population depends upon the surrounding forests for collecting dry twigs and for cutting green trees to meet their demand for cooking without realizing the hazards of deforestation. Firewood is transported from place to place by Bullock Carts and sold in the nearby towns.

Besides wood, there are other sources of energy which are being used by villagers for cooking; coal to a limited extent, dry bagasse for raw sugar making and cow-dung cakes for domestic usage.

Yet another important energy need of rural areas is for food production. Here again the villagers mainly use human and animal muscle power for tilling of land, sowing, reaping and for post harvesting needs. These techniques are used in conjunction with the direct utilisation of sunshine for post-harvesting processes.

In rural India, the common mode of transportation is the Bullock Cart. The tonnage of materials moved from place to place equals or exceeds the total haulage by motorised road transport and the vast railway network. The Bullock Cart is designed, fabricated and repaired in the rural areas itself with the simplest available tools. There has been a changing trend over the past decade or so to improve its performance by using second-hand motor vehicle tyres instead of the conventional wooden wheel with steel rim. Energy for manufacture and maintenance of improved designs is needed for rural areas.

During the crop sowing and harvesting period, the villagers are over-worked. At other times, there is tremendous manpower available for alternative productive output. This manpower could be used effectively for rural based industries. Energy inputs are vital for such an Integrated Village Development Programme to create employment opportunities.

The rural scene has changed over the past one or two decades in some of the areas. One can see planned housing, tractors, motor vehicles, electrically operated tube-wells and medical clinics. These changes, though a welcome feature, are the result of the provision of conventional and commercial forms of energy (electricity, oil) which are however, so expensive that only a small part of the rural community can benefit from them.

One of the new sources of energy which has gained considerable impact on rural life is the production of biogas and good manure as by-product from cowdung. Cowdung was hitherto being used inefficiently for making fuel cakes to provide thermal energy for cooking. Besides this, one occassionally comes across windmills for pumping water and for other rural mechanical energy needs like oil expellers and crop threshing.

The Research and Development for non-conventional energy systems is of crucial importance. In order to achieve the maximum advantage from the R & D programmes, there is an urgent need to assess the total energy requirements of villages and to design Rural Energy Package Systems to meet different needs in a locally optimised manner. Such systems will be a combination of solar energy, wind power and biogas to provide electricity, cooling, hot water and other thermal energy needs for cooking, washing, handloom cloth dyeing, sericulture etc. A close coupling of R & D activities in the Laboratories and Manufacturing units, with the actual needs and requirements of the rural community is vital for the success of this concept.

The problems are numerous and challenging. There are no solutions as yet which can be widely introduced. But only a planned approach which can provide the rural energy needs without drastically changing the social structure of the rural areas can be effective in the shorter term. Once changes start, long term effects of economic progress will bring out gradual changes of social structure without creating excessive turmoil. Therefore the technological solutions of today are to be considered in a limited time frame required for the first phase of Integrated Rural Development.

C. SOME CONTRIBUTED PAPERS FROM THE PARTICIPANTS

Statement

by

J.P. SEKA
UNCSTD Conference

The Secretary-General of UNCSTD (United Nations Conference on Science and Technology for Development) has asked me to thank publicly the CEC (Commission of the European Communities) for the interest and support which it has lent and continues to lend to the preparation of this conference. I also have the pleasant duty of thanking here Mr. Schuster and other officials of the CEC for their cooperation.

The Secretary-General of UNCSTD also wishes the organizers of ICSED the greatest succes in their work which began here this morning and will continue over the next few days. He considers this work, and the results expected from it, to be of the greatest importance.

UNCSTD, as you know, is based on the principles of the New International Economic Order. It aims to establish a climate of cooperation between the developed and the developing countries, touching on such important areas as:

1. The sharing of knowledge and acquired experience amongst all members of the international community;

2. Improving the ability to draw up a general scientific and technical policy in relation to development and planning;

3. The transfer of techniques for application to development;

4. Improving the countries' own capabilities, with a view to national autonomy;

5. Increasing collective autonomy by cooperation among developing countries;

6. Increasing the role of the UN in the field of scientific and technical cooperation.

In order to determine the problem and tackle it practically and effectively, the Secretary-General of UNCSTD used what he would call the ascending process, trying never to lose sight of the essential link which exists between science and technology and the political attitude of governments. Up to now, the Conference secretariat has received and analysed the contributions of over 120 governments and the five regional commissions.

We should note that Energy, the subject of ICSED, is one of the major fields of development chosen by the Preparatory Committee of UNCSTD as an illustration in national and regional documents submitted to the Conference Secretariat.

Although UNCSTD has deliberately been pitched at governmental level, the greatest importance is attached to the contributions of scientific groups on a national as well as regional and international level. Over recent months and during the few months left before UNCSTD, numerous meetings of experts and specialists have been and will continue to be held as part of the UNCSTD preparatory work. The ICSED is one of those meetings.

The governmental, intergovernmental and non-governmental form of participation used in the preparation of UNCSTD has meant that the preparatory phase of the Conference has been as important as the final phase, which will take place in Vienna (Austria) on 20-31 August. It could be said that UNCSTD started in January 1977 when its Secretary-General was elected; and an event such as ICSED represents another step, a major contribution to this long and complex process which we hope will lead to the adoption of a programme of action at Vienna.

We should like to thank the organizers for having invited us to take part in ICSED, which could make one of the greatest contributions to UNCSTD.

World Meteorological Organization and its Activities Related to Solar Energy

S. JOVICIC, WMO, Geneva

Mr. Chairman, Ladies and Gentlemen,

On behalf of the Secretary-General of the World Meteorological Organization, let me first congratulate the organizers and hosts of the conference for their efforts and the work which they have done to organize this interesting and useful event.

My understanding is that one of the main purposes of the conference is to improve the relations between all those interested in the use of solar energy in various parts of the world and I wish you to succeed in this work and to make progress in finding solutions which will be fair to all concerned.

My Organization is a World Organization and it is a specialized agency of the United Nations; its raison d'être as well as its approach to the problems is somewhat different and I consider it is my duty to try to inform you about the organization of WMO and some of its activities which are related to the subject of this conference.

Within the system of the United Nations, WMO was established after the second world war and, in fact, it inherited its basic responsibilities from the International Meteorological Organization which had started its work in 1873. Today, WMO includes the membership of 149 countries throughout the world. For those of you who are not aware of the system, I wish to indicate that in each of the WMO Member countries, there is a national Meteorological Service dealing, among other things, with networks of observational meteorological stations on its territory. These stations function according to WMO standards and in this way, the data resulting from the measurements and observations of meteorological elements are comparable between them in time and space all over the world. The meteorological elements include atmospheric pressure, temperature and humidity of the air, wind, clouds, precipitation, sunshine duration, solar radiation, visibility and various meteorological phenomena. Distributed on the surface of the Earth, there are on a global scale about 10 000 principal climatological stations, 40 000 ordinary climatological stations and 140 000 precipitation stations. Some of these stations started with their work two centuries ago and they have been operating continuously since that time. On the global scale, a considerable improvement in the distribution and quality of the stations was introduced after the second world war and, in many newly independent countries, which are becoming increasingly aware of the importance of the proper application of the information on climatic conditions to various sectors of their economy, Meteorological Services are making efforts to meet requirements of users of the climatological information and knowledge.

In connexion with energy problems, there is a WMO Plan of Action which was approved in 1976 and within which the highest priority is allocated to solar and wind energy. At this moment, I have to mention that according to this Plan of Action, the solar energy does not include wind energy and, therefore, the two areas are considered separately.

For solar energy, I am glad to be able to inform you that, in last October, WMO organized in Geneva a five-day meeting between meteorologists and technologists dealing with the problem. It is very pleasant to realize that several of the participants of the WMO meeting are attending the present conference in Varese.

You may wish to note some of the conclusions of the WMO Solar Energy Meeting, which are reviewed as follows.

The most important conclusion concerns the preparation of a WMO Technical Note on Meteorological Aspects of the Utilization of Solar Radiation as an Energy Source. The main purpose of the Technical Note is to give to both engineers and meteorologists all over the world and in particular in developing countries, the guidance and advice on how to use, for solar energy purposes, meteorological information including solar radiation data. Taking into account the importance of this Technical Note, especially for developing countries, the possibilities of using solar radiation measurements for solar energy purposes will be indicated in its introduction. However, in the areas where reliable solar radiation data are not yet available, the advice will be given to use in the meantime the other meteorological information, such as data on sunshine duration and cloud amount as well as satellite information. Such assessments will be recommended as a temporary solution only and every effort should be made to introduce real measurements of solar radiation with a view to using them to overcome the existing difficulties. The needs to improve the networks of solar radiation stations, i.e. to increase their number as well as to complement the programme and improve the quality of their measurements, will be stressed adequately.

In connexion with the results of solar radiation measurements, Meteorological Services are advised:

(a) to archive these data as detailed as possible;

(b) to publish on a regular basis reliable data, and

(c) to send regularly requested solar radiation data including appropriate information on calibration of instruments and quality control, for publication in the World Bulletin on solar radiation and radiation balance data.

Furthermore, the subject of WMO Resolution 15 (EC-XXX)—Inventories of climatological stations and catalogues of climatological data, covers as appropriate, radiation stations and radiation data too.

Within the WMO activities, it is planned to establish for each month world climatic maps on solar radiation. Because of the gaps in space and time of solar radiation data, it will be necessary to prepare maps on sunshine duration and daily cloud amount which in combination with available solar radiation data, will enable meteorologists to prepare world maps on solar radiation.

Similarly to the 1978 Solar Energy Meeting, WMO is organizing the holding in 1979 of a meeting on wind energy.

On 30 April 1979 the Eighth World Meteorological Congress will begin in Geneva, to which 149 Member countries will send their delegates. A special item on the Congress programme is devoted to energy problems and the decisions of the Congress will be carried out by WMO and national Meteorological Services in 149 countries.

It may be of interest to mention some aspects of the machinery of WMO. There is a system of eight WMO Technical Commissions on which Member countries are represented by their experts. These commissions have their working groups and rapporteurs dealing with specific subjects. One of the commissions is the Commission for Special Applications of Meteorology and Climatology (CoSAMC). It meets every four years and in view of its achievements and new developments, establishes its programme for the following four-year period. CoSAMC includes working groups on energy problems, human settlements, climatology, building climatology, climatic atlases, guidance material, air pollution, human biometeorology, tourism etc.

There are also six Regional Associations of WMO covering six geographical areas—of Africa, Asia, South America, North and Central America, South-West Pacific and Europe. The Regional Associations have their working groups and rapporteurs which are assigned to consider regional aspects of technical problems carried out centrally by the system of technical commissions.

I hope that in the further activities associated with the outcome of the present conference, the existence of the World Meteorological Organization and the availability of its achievements will be duly taken into account with a view to coordinating the future efforts related to the use of solar energy in various parts of the world.

Thank you for your attention.

Thank you, Mr. Chairman.

Statement

by

Prof. M.A. ABBAS
Federation of Arab Councils for Scientific Research

Mr. Chairman, Participants,

On behalf of the Secretary-General and all members of the Federation of the Arab Councils for Scientific Research, I would like to express my deep thanks for the kind invitation given to the Federation to participate at this conference.

As you know Arab Countries have plenty of solar energy which, on average means having a sunshine duration amounting to about 3 400 hrs/a year.

There are many activities in the utilization of solar energy going on in different Arab states. These include:

1 — Water desalination in Jordan, Iraq, Sudan and Saudi Arabia.

2 — Solar concentration for power generation in Egypt.

3 — Heating and cooling in Kuwait.

4 — Flat plate collector projects in Iraq, Sudan and Egypt.

5 — Solar drying in Libya and Sudan.

The General Secretariat of the Federation is looking after these activities and gives them great support. It has also been planned to undertake joint research projects in different parts of the Arab States.

It is of interest here to say that we do have the human resources and the essential facilities to run the above mentioned projects, but we have a great desire and hope very sincerely to cooperate with the Commission of the European Communities, and any other organization in the world, side by side as equal partner for the benefit of the co-operating countries and for the service of humanity.

Again I would like to thank all those who prepared and organised this conference for making this opportunity available for us to participate.

Wishing this conference all success.

The Social Price of Technology Transfer

by

C. TRINDADE, Brasil

Although we may employ the same language for communication (e.g. English and French in this conference) we listen according to our cultural filters. This is the basis for the lack of understanding in international relations and even among developing countries sharing common problems. Consequently the issue of perception is at the root of understanding the process of technology transfer.

In this context, I would like to examine the concept of the true price of technology as perceived by both seller and buyer. Before getting to that I would like to point out that we can approach the problem of technology transfer from at least two viewpoints, namely, the efficacy and efficiency of the process.

The efficacy is the least discussed aspect of the problem. To me efficacy means the adequacy of cultural and physical resources to the technology at hand.

The efficiency reflects the extent to which essential knowledge and skills related to the technology at hand are absorbed. The evidence of true absorption is the ability of the recipient country (firm) to sell back the technology received, upgraded, to the original supplier (or equivalent). In the conventional economical analysis the technology issue is usually considered of second priority because the financial price paid for technology transferred is usually expressed as a percentage of the selling price and that seldom reaches over 5%.

Consequently, it is necessary to define a new way to express the price of technology transferred in a way to reflect the perceptions of both seller and buyer.

Therefore, in the context of efficiency of technology transfer it is useful to define the following "price" of technology, namely:

$$P = \frac{\$}{Q} = \frac{\text{Price paid for technology}}{\text{Fraction of essential knowledge and skills absorbed in the transfer process}}$$

If we take a given specific technology say T_1 then $\$$ would be constant. In this case if we plot P versus Q we get a hyperbolic curve.

Looking at this curve shown in Fig. 1, it becomes clear that if the receiving country (firm) has a low Q, i.e. limited human resources qualified to absorb the technology at hand, then it should perceive itself paying a very high price "P". At the same time the seller perceives always the same selling price, at $Q = 1.0$, because the seller has a very qualified human resource base. This is key to understand the difficulties of dialogue between sellers and buyers of technology. It also emphasizes the key role of developing an adequate human resource basis in the recipient country, firm or institution.

It should be clear that it is primarily up to developing countries to find out for themselves what they want to accomplish in solar energy applications. The European Community can certainly help but only within the context of clearly defined programs reflecting objectives set by the developing countries.

To exemplify the above statement using the concept shown in fig. 1, a country (or firm or institution) can start transferring at a high price "P" (equivalent to low "Q") a certain solar technology. Simultaneously the country (firm or institution) could start an accelerated program of increasing "Q" in such a way as to cut down the perceived price "P" paid in successive transfers. That requires a clear decision on the part of the receiver which imply political action or decision.

Alternatively, the country (firm or institution) could, for a specific technology, get its "Q" close to 1.0 before engaging in technology transfer.

Finally, it occurs to me that all developing countries have armies of University students abroad and at home. If a political consensus can be obtained in each of our countries to incentivate some of these people to work, abroad and at home, on research problems of interest to solar energy applications, we would have accomplished an increase in the specific "Q" at a low cost. Furthermore, we would have prepared the environment at home for the reentry into their countries from abroad. This is not easy to accomplish but it seems worthwhile trying.

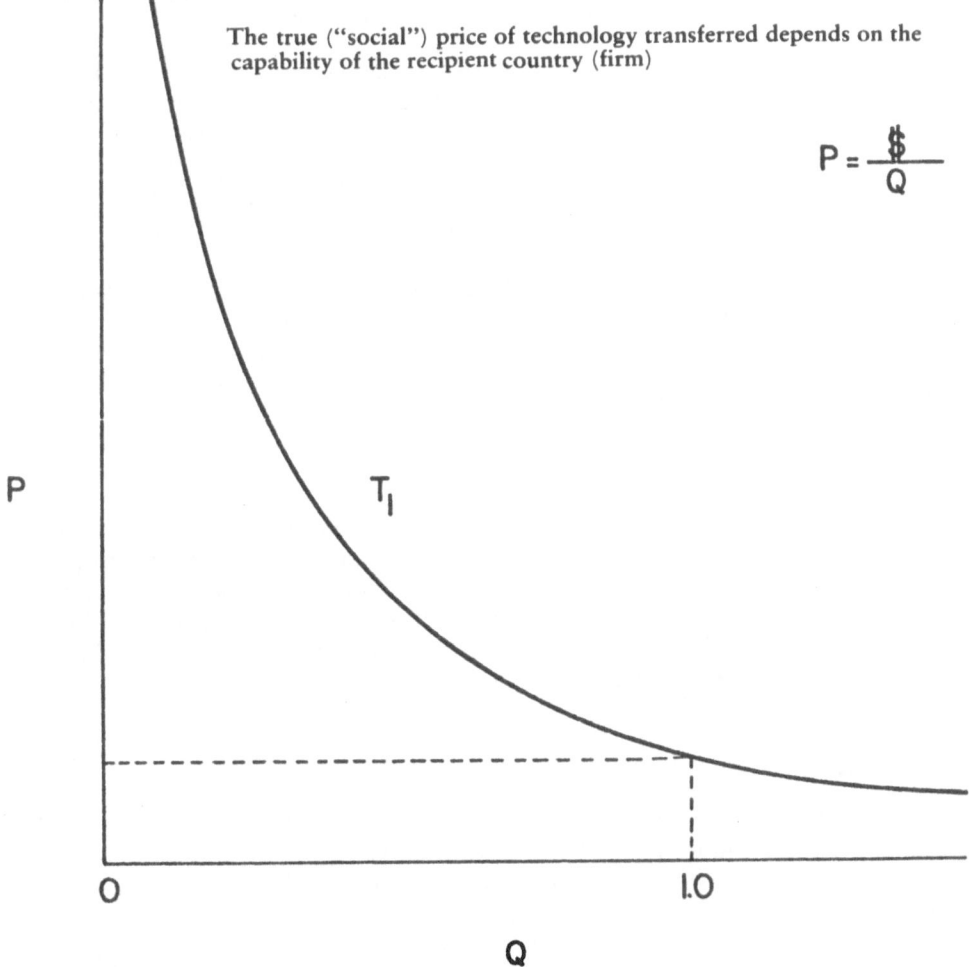

The true ("social") price of technology transferred depends on the capability of the recipient country (firm)

$$P = \frac{\$}{Q}$$

T_1

P

O   1.0

Q

Q: Fraction of essential knowledge and skills necessary to absorb technology.

Industrialization, Technology Transfer

by

D.M. KROKO
Ivory Coast

Some of the statements made by previous speakers on the subject of technology transfer remind me of the attitudes so often taken by some representatives of the developed countries, for whom, as we heard yesterday in this very room, this transfer — whatever its real meaning and precise content — is a gesture of international charity by means of which the industrialized countries come to the aid of the developing countries. This leads me inevitably to think of that wise proverb which says something like: "To give a fish to a hungry man is good, but to teach him to fish is infinitely better." This proverb clearly shows that the idea of technology transfer as a charitable act should be rejected, as it is inaccurate: true technology transfer should maintain the human dignity of both parties.

Too often as well we get the strong impression that those whom we should like to think of as equal technological and technical partners see our countries merely as markets to be conquered, or already conquered, which they strive by all possible means to make their own preserves and to exploit with no thought for either socio-economic conditions or local interests. Experience shows that this point of view leads inevitably to the importation of foreign technology which is often sophisticated, usually ill-suited and invariably inaccessible to nationals, as it leaves no room for training of management, engineers and scientists capable of assimilating it. Nor, for this reason, could such a view be acceptable to the LDCs for they cannot allow themselves to be satisfied with producing raw materials, whether it be cocoa beans, timber or data on solar radiation, and then looking to the industrialized countries for the finished products they go to make, such as chocolate, furniture or pumps and other solar equipment.

Therefore it seems imperative and urgent to rethink the very concept of technology transfer, in the context of open and straightforward international cooperation. The field of solar energy, by its very newness, presents an exceptional opportunity, as neither the industrialized countries nor the developing countries have yet totally mastered it. Consequently, this Conference will be productive only in so far as it goes beyond endless declarations of intent and provides an approach which will ensure effective international cooperation.

For my part, I should like to suggest that priority be given to close cooperation in the training of local skilled personnel, starting with the schools.

In my view, this is an essential step. The reasons are many and obvious: the adaptation of solar techniques to the life-style of the local populace, technological independence, etc. In this context, the following action could be taken at once:

1. The industrialized countries should make a special effort to make their specialized schools more accessible to LDC nationals. This does not mean that the level of the schools' entrance examinations should be lowered, nor that a

special category of students should be created, but that the number of students the schools can accept should be increased, taking into consideration the needs of these countries.

2. At the same time, specialised schools should continue to be set up in the LDCs, both at national and sub-regional level. Examples of this exist or are being built, such as the UPDEA (Union des Producteurs, Transporteurs et Distributeurs d'Energie Electrique en Afrique) Engineering school at Abidjan. A sustained effort should be make in this direction, the industrialized countries supplying the necessary teachers as well as the various items of plant and laboratory equipment, within bilateral and multilateral agreements.

I feel that such a foundation would be likely to ensure the success of industrial cooperation, with respect and dignity on both sides.

*

Market Penetration Issues Facing Solar Energy Alternatives

by

C. TRINDADE, Brasil

Energy issues are decided upon primarily on political grounds. Considerations relative to technology and conventional economics are important inputs to decision making, but in many practical situations political considerations override all other inputs. For example, the Brasilian alcohol program would not exist if conventional economics were the dominant input in the decision that brought the program to life. In conventional economical terms it costs about 3-4 times more to produce fermentation alcohol than to buy petroleum. However, if one takes into account benefits such as foreign currency savings, intensification of rural activity, lessening of urban migration, better income distribution, security of supply etc. then it may be justified to make a decision which goes against the conventional economic analysis. However, to make such a decision requires political clout.

In essence, if a given country wants to implement solar energy alternatives against conventional economic evidence but with prospective positive social benefits, it can only accomplish its goal via political decision. In other words, early introduction of potentially and socially beneficial solar energy alternatives requires political clout, provided the technology does not constitute a barrier in itself.

In the above context, I would like to see this conference address all the issues before it from the angle of market penetration, in additional to technical and economical considerations.

If we try to analyse prospective market penetration of new energy technologies, we might be able to locate key barriers and therefore plot adequate strategies.

It is well known that market penetration of any new technology faces many barriers such as technological, economic, social, political, psychological etc.

It is also well known that in most cases, when the technology is relatively developed, the key barriers are non-technical. They tend to be social, political and psychological. For example nuclear energy offers no technical and in most cases no economical difficulties, but due to psychological barriers the American and European nuclear energy program has almost come to a halt in recent years.

These concepts with all their qualifications, can be very helpful in plotting strategies for the market penetration of new solar energy alternatives. Currently, in this field the barriers to market penetration are technological in most cases. When that is clear then the strategy should include R & D to remove uncertainties in the technology. However, there are many cases, as exemplified by the Brasilian alcohol problem, when the barriers are definetively non-technical. In such cases political decision is required to accomplish market penetration.

Synthesis of the Solar Energy Program

by

G. Rodriguez ELIZARRARÁS
Latin American Organisation for Energy (OLADE)

OLADE, within its promotion, coordination and orientation function in the field of energy in Latin America, has structurated a solar program starting from the local frame of activities. The final objective is the full utilization of this energy source at a technological level parallel to the developed countries with a step-by-step implantation of human and natural resources available in Latin America. The purpose of the program is to cooperate to achieve a great degree of hydrocarbon independence, towards a general improvement in the wellbeing of our countries.

In the solar development context, Latin America presents a diversified and heterogeneous panorama. The development programs in the countries are in a relative state of initiation, with some exceptions. In a constructive structuration towards the final objective, is OLADE's Permanent Secretariat criterion, to formulate a step-by-step leveling program starting with an immediate implementation which consists of two actions: in one way, to establish and maintain an informative homogeneity of current knowledge, the basis of energy development, to make possible the Latin American technician's professional development and to obtain assistance of international experts. And in another way, start a continual utilization of solar energy from its current and regional state, elaborating pilot projects with the necessary flexibility that makes possible an immediate adaptation for its direct application in different localities.

For the establishment of these immediate actions, OLADE sets up a directive body, nominated Group of Solar Coordination (GCS), to connect and coordinate basic centers to be created in different countries.

These centers have the following nominations and functions:

1. Center of Documentation, Information and Diffusion (CEDID) whose function is to concentrate, process and diffuse information.
2. Center of Climatology (CECLI) to collect, analyze and process climatologic data.
3. Center of Technological Exchange (CITEC) to facilitate the technological exchange and the formation and training of technical personnel.
4. Center of Project Implementation (CIPRO), whose main objectives are the determination and implementation of particular projects.

This basic structure will guarantee the most broad coordination during the transition stage to the full level of world technology, in which OLADE promotes and receives the necessary technical transfers, both internal and external, to canalize them towards an optimization in the efficiency of its uses.

Latin America is particularly an agricultural region, and within the classification of rural, urban and industrial needs, the first demand a prioritary attention within a limited scope of action.

Thus, OLADE has programmed pilot projects based on the production of biogas and the drying of agricultural and feeding products to be realized with the regional resources available, without considering that these constitute the total rural needs. Afterwards operational programs will be structured for the following equally important and immediate solar projects:

— Production of liquid and solid fuels
— Irrigation and pumping
— Electric production
— Water desalting
— Glass-house for the reduction of water consumption
— Environment and product refrigeration
— Environment and water heating.

Finally, due to the changing pattern of technical, economical and regional conditions, OLADE's Permanent Secretariat opinion is that for the moment it is too soon to structure a program of major reach. However, guidelines are being formulated for the future development based on a conditioning over the results obtained in this initial stage.

The Energy Crisis and the Potential for Utilization of Renewable Alternative Energy Sources in Developing Countries The Role of IDB Assistance to its Member Countries

by

L. da SILVA
Inter-American Development Bank

The energy crisis and the gradual depletion of petroleum resources create conditions in which governments of developed and developing countries alike will have to implement policies and measures to facilitate the transformation of the present oil-based economic structures into ones less dependent in the future on fossil fuels. A frontal attack on energy problems is required by all countries. A strong committment to well designed programs to increase the efficiency of energy use, to develop indigenous supplies from conventional and non-conventional energy sources and to mobilize the required technical and financial resources, all within the framework of adequate energy planning, are the elements necessary to elicit proper responses to the energy problem.

The agricultural sector plays an important role in the overall energy picture as both an important supplier and user of energy sources. The increases in oil prices and the rapid depletion of fossil fuels are likely to affect future agriculture growth prospects and improvement in the quality of life of the rural population. Present energy supply systems in rural areas are inadequate to satisfy basic human needs and to reduce the inequalities between the rural and urban population, as well as between the various income groups within the rural areas. In effect, only the high income groups in rural areas seem to have access to commercial energy supplies and are thus in a better position to increase production, productivity and income. In all these cases, the present rural energy supply structure is inconsistent with broader socio-economic development objectives and inherently unsustainable in view of the pressures from rising population and the higher costs of commercial fuels. Attempts to increase employment opportunities, income and the quality of life in rural areas through a better provision of potable water, sanitation services, education and shelter require significant amounts of additional energy and modifications in the energy supply and demand structure.

In the energy sector, as in all other sectors, the most acceptable solutions are those which are easy to implement, are consistent with the existing institutional framework and avoid to the greatest extent, possible disruptions to the established set of values and procedures. Satisfaction of energy requirements within this framework is normally sought by means of projects designed to increase or maintain accessibility to fossil fuels and to the conventional sources of energy, mostly in the form of electricity from large scale central stations. While increased amounts of energy will have to be provided for the rural areas, widespread acceptance and strict adherence to this position, however, are not likely to lead in all cases to a permanent solution to the long-term energy supply problems of the lowest income group.

Within this overall context, the need to provide economical solutions to the energy problem in rural areas can be central for improving the living conditions and for providing meaningful employment opportunities to a large segment of the population living in substandard conditions. There is a general recognition that the renewable alternative energy sources and especially certain forms of solar energy, can play in important role in rural settings.

But if alternatives to fossil fuels are to play such a role, greater efforts and long-term commitments will be required by both local institutions and the international community in development, transfer, adaptation and dissemination of renewable alternative sources technologies. These needs, however, are normally received with skepticism and resistance as they do not conform to the "traditional ways of doing things", in addition to mental inertia and to the widespread perception in many quarters that only large scale and capital intensive technologies are desirable. Recognizing that there are a variety of difficult problems to overcome and that these present themselves as great challenges to alle parties; by the same token they also present great opportunities for assistance with, most likely, high payoffs.

The opportunities rest on the ability to induce a redirection of the "traditional energy thinking" and in bringing together programs that can effectively and efficiently promote the utilization of technologies that better reflect factor endowments and the set of problems in developing countries. In effect, the development and utilization of renewable alternative energy sources can only be considered within the context of strongly interrelated socioeconomic and technical variables. The traditional economic criteria (cost efficiency) used to determine the feasibility of large scale, capital intensive energy projects has to be reviewed in order to take into consideration other objectives and the realities confronting our member countries, especially those of a great part of the population with extremely low living standards and with poor employment prospects. This is not tantamount to say that the development of renewable alternative energy sources is a panacea, even for the requirements of the low income groups in rural areas. Unquestionably, present solar energy technologies are faced with several constraints, the seriousness of which varies and depends on the particular set of circumstances and specific applications for which they are intended on a country by country basis.

As indicated previously, with the advent of the energy crisis there has been a great deal of discussion on the potential uses of solar energy to meet energy requirements in both developed and developing countries. Presently, there seems to be no major technical barriers for the utilization of solar energy systems—most of the main obstacles preventing their more intensive use include the following: (a) high investment costs in relation to other alternative sources; (b) lack of adequate knowledge about the alternative solar energy technologies presently available and the anticipated technogical developments in solar conversion; (c) insufficient information on the economic benefits and costs associated with the use of solar energy systems and with the many technical options; (d) inadequate awareness of the market potential and requirements of different end-uses; (e) difficulty of implementing sociocultural and institutial transformations that may be needed as prerequisites for economic utilization of these technologies; and (f) limited commercial availability of solar equipments.

The shortcomings or deficiencies of any of the new energy technologies could be at least partially eliminated in an integrated system composed by several subsystems. In these integrated systems the overall yield can be significantly increased and the overall cost per unit of energy production reduced. In this respect, it should be noted that the discussions of the rural energy supply options are almost exclusively framed in terms of small scale technologies, benefiting a limited number of consumers per system. While the impact of these small scale technologies can be significant at the consumer level, especially to the lowest income group, at a national level their marginal contribution to increased and diversified energy supplies is relatively small over the short and medium run. As large energy quantities are required, larger scale technologies become more attractive and conceivably, the only feasible alternative, from both a technical and economic point of view. It is, however, questionable that the options to be provided by the renewable alternative energy sources can in effect solve the energy problems faced by the rural population. Not only is their potential contribution to energy supplies marginal but also their dissemination is frought with great difficulties emerging from existing social and cultural constraints, as well as from inadequate institutional framework and marketing system just to mention a few.

It is generally accepted, notwithstanding, that small and medium scale solar energy systems, with unsophisticated technology and low maintenance requirements can be an attractive option for many rural communities in the developing countries, especially in remote areas with costly access to conventional forms of energy and with their main supply of energy being derived from firewood. There are no reliable data on the actual composition and levels of utilization of these forms of energy, but it has been observed that, in general, they represent a very high proportion of total energy consumption and in certain cases they exceed the amount of reported commercial energy conmsumption. In effect, in spite of the gains from the implementation of ambitious rural electrification programs, significant proportions of the rural population still relies heavily, if not exclusively, on the supplies of traditional fuels to satisfy some of the most basic necessities of life such as cooking, heating, and lighting.

Several surveys indicate that cooking alone accounts for the greater proportion of the energy consumption in rural ares. Since this demand in inelastic and fuel substitution options are very limited, the rapid depletion of traditional fuels, and the consequent increases in costs in obtaining them, has in addition to environmental considerations, serious impact on the struggle for subsistence in the low income groups. If an increase in the efficiency in the rate of conversion of these traditional sources of energy could be obtained (as for example by introducing more efficient stoves or by using solar cookers, whenever possible), the pressure on forest resources and land holding capacity would decrease and more labor would be freed for productive activities.

In recent years there has been a substantial and indiscriminate use of firewood and agricultural wastes as these sources of energy are substituting higher cost petroleum-based products, such as kerosene, among the lower income groups and in cottage industries as well. Higher rates of utilization of these traditional sources of energy are rapidly depleting forest resources and causing severe damage to the environment through land erosion and loss of soil fertility. This situation threatens to reduce the possibilities for further and necessary gains in agricultural production—especially in subsistence-type agriculture—and in the quality of life in many areas of the region where the massive depletion of these resources tends to disrupt the life-supporting ecosystem. It is indeed questionable that the energy requirement necessary for additional employment opportunities and for improving the quality of life will be available at reasonable and acceptable cost levels for a significant proportion of the population since at the present and expected rate of utilization the long-term prospects for the traditional fuels is no more encouraging than that for oil. It can be said, in effect, that the increased costs of energy-based fertilizers, pesticides, herbicides, and other agricultural inputs which made possible the development of the high yielding varieties of the Green Revolution, and the dangers to the environment from the indiscriminate uses of traditional fuels, are among the major problems confronting all those countries struggling to increase food production while diversifying at the same time the present structure of energy supplies.

There is, evidently, a very close relationship between the energy crisis as it is usually defined and the other, less perceived crisis stemming from the indiscriminate use of forest resources and rural wastes as sources of energy. This link stems from the emergence of new forms and higher levels of energy consumption as substitute for the traditional patterns existing in rural and urban-poor areas, which as indicated has been one of the main elements accounting for the substantial gains in productivity throughout the time. Energy and development, therefore, have to be viewed as an interrelated process.

IDB Assistance to its Member Countries in the Field of Renewable Alternative Energy Resources

Within the framework of its overall objective to foster economic and social development of its member countries, either on an individual or on a regional basis, the Inter-American Development Bank has recognized the importance of encouraging the utilization of renewable energy resources to meet the requirements of both urban and rural population. At the request of its member countries, the IDB is in a position to support high priority projects in this field either through: (a) technical assistance operations for preinvestment studies (including resource assessment, identification of investment opportunities, prefeasibility and feasibility studies and engineering and final designs), technology transfer, development and adaptation, institution creation and/or strengthening, as well as sector planning, policy formulation and investment programming; (b) financing for high priority and viable capital investment projects or programs formulated according to national or regional socio-economic development plans and objectives.

Past and Present Operations

IDB's program in small scale renewable sources of energy at the present time, however, is modest and mainly concentrated on a non-reimbursable assistance to the Instituto Centroamericano de Investigación y Tecnología Industrial's efforts to

135

develop intermediate technologies for rural areas in the Central American countries (see Annex 2); and on the design of an applied investigation program for the utilization of solar energy in the rural areas of the Dominican Republic. The first of these activities, which is part of a much larger program, envisages the development and dissemination of small and medium scale solar crop dryers and gas production systems from agricultural and animal wastes. The program with the Diminican Republic reflects this country's request for IDB assistance in the development of pilot projects related to the utilization of solar energy (thermal and photovoltaic) in several high priority rural uses. At the present time, the IDB has approved an initial technical cooperative operation for the Instituto Dominicano de Tecnología (INDOTEC), designed to define the scope of this proposed assistance and to select the most appropriate technologies for further experimentation, adaptation and dissemination. In addition, the IDB has commissioned a short-term study on existing patterns of industrial molasses utilization in several of the Caribbean countries and territories and to identify alternative industrial uses, including the production of power alcohol as a substitute for gasoline. This survey was conducted under the aegis of the Caribbean Development Bank and was completed in early 1978.

Another activity of the IDB in this field is a non-reimbursable technical assistance to the Nicaraguan Istituto de Fomento Nacional (INFONAC) approved in 1975 for a study to determine the feasibility of alcohol production from sugar cane juice and molasses. Although this project is essentially geared to the alternatives for substituting alcohol for petroleum in chemical process, the potentials for power alcohol production will also be identified.

In the field of capital project financing, several of IDB global loans for agricultural and livestock development as well as rural development programs are likely to incorporate the financing of investments in renewable small scale sources of energy. This is the case, for example, of a dairy program in Guatemala, in which electricity is being provided by micro hydro stations, water pumping through windmills and water lift for irrigation through pumps working exclusively on the basis of water pressure.

Outlook

To the extent to which major lessons are learned regarding the applicability of these technologies and wide commercial applications are fully developed, the participation of IDB investment financing and technical assistance programs in countries in which major opportunities exist will undoubtedly become more significant in the years ahead.

The high priority given to rural development and to improve the quality of life of the low income groups given by the IDB's Board of Governors provide, in effect, the framework for the promotion of renewable alternative energy sources in our member countries.

In this context, the IDB policy on the use of intermediate technologies (see Annexe 3) provides a framework for supporting the development and dissemination of renewable alternative energy sources in its member countries, while the cross-fertilization of experiences from the on-going efforts in the region's research institutions can be fostered by the Bank's Intraregional Technical Cooperation Program.

Reported Solar Energy Investigations in Latin America
(Partial Listing)

Country	Institution/Organization	Field of Activities and Investigations
Argentina	Observatorio Nacional de Física Cósmica	Thermal conversion, distillation, fluid heating, photovoltaic cells
	Universidad Nacional de Salta	Solar dryers, hot water heaters, and solar ponds
	Universidad Nacional de San Luis	Distillation and desalination
	Universidad Nacional de Rosario	Passive systems for buildings
	Universidad Nacional de la Pampa	Distillation and solar cookers
	Comisión Nacional de Estudios Geo-Heliofísicos	Various solar technologies
	Observatorio Nacional	Solar radiation measures
	Universidad de Mendoza	Passive systems for buildings, solar heating and cooling
	Universidad de Córdoba	Solar radiation and insolation
Bolivia	Laboratorio de Física Cósmica de Chacaltaya	Water heaters and other solar applications
	Universidad Mayor de San Andrés	Various solar technologies
Brazil	Universidad Nacional de Paraíba	Distillation, fruit dryer, solar cookers, solar pump for irrigation, solar radiation measures
	Instituto Tecnológico de Aeronáutica	Desalination, cooling systems
	Centro de Pesquisas de Cacau	Crop dryers, heat storage, passive energy systems and solar cells
	Comisão Nacional de Alcool	Alcohol production
	Universidad de Campinas	Flat plate collectors, solar crop dryers, refrigeration and solar architecture
	Instituto de Pesquiza de Marinha	Methane from water hyacinths digestors, water distillation, ocean thermal differences, wind energy
	Departamento Nacional de Meteorología, Ministerio de Agricultura	Solar radiation and insolation

137

Country	Institutions/Organization	Field of Activities and Investigations
Chile	Universidad del Norte	Solar radiation measures, hot water heaters, solar collectors, fish drying, distillation, and solar architecture
	Universidad de Chile	Solar cooling
	Universidad Técnica Federico Santa María	Solar radiation, solar cookers, hot water heaters, greenhouses, and solar water pumps and desalination
	Comisión de Energía Solar	Solar thermal electric generation plants
	Universidad Técnica	Solar collectors and water treatment
Colombia	Instituto de Investigaciones Tecnológicas	Hot water heaters, solar cookers and cooling and solar crop dryers
	Universidad Nacional	Solar crop dryers
Costa Rica	Universidad de Costa Rica	Hot water heaters, cooling wind turbines and gas production
	Universidad Nacional	Solar seed and crop dryer
Dominican Republic	Universidad Católica Madre y Maestra	Solar radiation measures, solar cookers and hot water heaters
	Universidad Autónoma de Santo Domingo	Various solar technologies
	Consejo Estatal del Azúcar	Gas production from livestock waste
Ecuador	Programa para el Desarrollo Regional del Sur del Ecuador	Biogas production
	Escuela Politécnica	Hot water heaters
Guatemala	Universidad de San Carlos	Gas production from urban and agricultural wastes
	Investigaciones Científicas Asociadas del Altiplano	Gas production from livestock wastes and solar crop dryer
Honduras	Corporación Hondureña del Banano	Solar dryer
Jamaica	Ministry of Mines and Natural Resources	Hot water heaters, solar collectors, solar seed dryer
Mexico	Universidad Nacional Autónoma de México	Solar radiation measures
	Centro de Investigación de Materiales	Solar cooling, solar collectors and photovoltaic cells
	Instituto Politécnico Nacional	Solar water heaters and desalination
	Instituto Tecnológico y de Estudios Superiores de Monterrey	Solar architecture, heating and cooling

Country	Institution/Organization	Field of Activities and Investigations
Peru	Instituto de Investigaciones de Aplicaciones de la Energía Solar	Solar water heater, solar collectors and solar seed dryer
	Investigación Tecnológica, Industrial y de Normas Técnicas	Various solar technologies
Trinidad & Tobago	University of West Indies	Solar water heaters, solar seed dryer, solar cooling and photovoltaic cells
Uruguay	Departamento de Energía	Wind and energy for water pumping and electricity generation
Venezuela	Instituto de Investigaciones Científicas	Photovoltaic cells and several solar technologies
Regional	Instituto Centroamericano de Investigación y Tecnología Industrial (ICAITI)	Solar crop dryer, biogas, alcohol production

IDB/ICAITI. Program on Intermediate Technologies (ATN/SF-1605)

Project 1. Fuel Production from Agricultural and Animal Wastes

1. Background

Significant proportion of the rural population in the Central American countries relies heavily on fuelwood to satisfy energy requirements. Indiscriminate uses of firewood are rapidly depleting forest resources and causing severe damage to environment through land erosion and loss of soil fertility.

2. Project Objectives

a. Selection, adaptation, construction, field testing, and demonstration of anaerobic digestor(s) for fuel production (methane gas) based on agricultural and animal wastes.

b. Dissemination of technology.

3. Location

a. Technology selection and adaptation; Guatemala, ICAITI.

b. Prototype construction and demonstration; Costa Rica, COOPECORONADO, R.L., dairy cooperative (110 cooperative members).

c. Dissemination: (tentative).

   El Salvador. (Centro Nacional de Tecnología Agropecuaria, CENTA, and Ministerio de Agricultura y Ganadería);

   Costa Rica. (Universidad de Costa Rica, Instituto de Fomento Cooperativo, Instituto Costarricense de Electricidad);

   Guatemala. (Banco de Desarrollo Agrícola, Programa de Ganado del Ministerio de Agricultura, PRODEGA);

   Honduras. (Dirección de Fomento Cooperativo);

   Nicaragua. (Instituto de Fomento Nacional).

4. Budget (Preliminary).

Total	$ 145,000.00	
of which:	$ 87,000.00	for construction, field testing, demonstration and dissemination.

5. Termination Date: mid 1980.

Project 2. Utilization of Solar Dryers dor Grain Preservation

1. Background

Production of basic grains (corn, beans, wheat, sorghum and rice) by small farmers are faced with humidity problems and high losses. Lack of adequate drying equipment at the farm level is also reflected in lower prices and, therefore, income for producers.

2. Project Objectives

a. Selection, adaptation, construction, field testing and demonstration of small and medium scale solar crop dryers.

b. Dissemination of technology.

3. Location

a. Technology selection and adaptation: Guatemala, ICAITI.

b. Prototype construction and demonstration: Guatemala, Cuna del Sol Cooperative.

c. Dissemination (tentative).

> *Guatemala.* (Centro Mesoamericano de Estudios sobre Tecnologías Aplicadas, Instituto de Ciencias y Tecnologías Agrícolas, Dirección General de Servicios Agrícolas);
>
> *Costa Rica.* (Universidad de Costa Rica, Instituto Nacional de Fomento Cooperativo);
>
> *El Salvador.* (Centro Nacional de Tecnología Agropecuaria, del Ministerio de Agricultura y Ganadería);
>
> *Honduras.* (Dirección de Fomento Cooperativo);
>
> *Nicaragua.* (Instituto de Fomento Nacional).

4. Budget (Preliminary).

Total	$ 110,000.00	
of which:	$ 58,300.00	for construction, field testing, demonstration and dissemination.

5. Termination Date: mid 1980.

Excerpts from IDB Policy on the Use of Intermediate or Light-Capital Technologies

General Policy and Goals

1. In the application of technology in projects financed by the Bank, the Bank will seek to ensure that the technology used is appropriate to development goals of the borrowing country or region and to local socio-economic conditions. The appropriateness of such technology would be assessed in terms of its impact on resources relative to local factor costs, the size and preferences of local markets, the skills of local labor and management, the availability of raw and semi-finished goods, and in relation to the present and future capacity for local planning, and implementation.

2. Capital- and energy-intensive technologies may be inappropriate for developing countries because they use too much scarce capital, too little labor or produce products less suitable for the existing markets. In many cases, intermediate or light-capital technologies would be more appropriate to achieve a country's development goals given a scarcity of capital and an abundance of relatively unskilled labor. 1) Use of these technologies to increase the productivity of the local labor force could produce an important additional source of the savings needed at a negligible investment cost, for self-sustained generation of capital which is a prerequisite for the long term growth of developing countries. Therefore, the Bank will make the development and use of intermediate or light-capital technologies an increasingly important facet of its programs. This should be done as soon as possible and be applied to all appropriate Bank activities, after case by case analysis.

3. The Bank will assist developing member countries to incorporate appropriate light capital or intermediate technologies in Bank financed projects through the project analysis process and in technical cooperation provided to those countries, seeking to collaborate with them in improving their capacity to adapt and develop such technologies. The Bank will also endeavor to stay abreast of the latest developments in intermediate or light capital technologies, including light-capital energy technology, and will assess, evaluate and promote innovations needed to make available to developing countries technologies appropriate to their needs.

Loans

4. In its development loans, the Bank will encourage the use of technology or combination of technologies which are most appropriate, bearing in mind local socio-economic conditions, availability of resources, need of products and the long-term objectives of development of the borrowing country in appropriate cases. The Bank will promote specific projects to cause the diffusion of intermediate or light-capital technology and measures to adapt capital-intensive technologies as a function of local needs.

5. While seeking to promote the use of intermediate or light-capital technologies in its development projects, the Bank recognizes that the use of capital-intensive technology might be more appropriate in certain circumstances, depending on the context of the specific project within the borrowing country's development program and the economic and social conditions in a particular locale.

---

¹) The terms intermediate or light-capital technologies refer to tools and techniques that are neither primitive nor capital intensive, but which are appropriate for a particular country at a particular time.

Technical Cooperation

6. Through its technical cooperation program, the Bank will endeavor to stimulate intermediate or light-capital technologies and seek through Bank financed projects to develop and enhance the capacity in the developing countries to adapt and develop intermediate or light-capital technologies and improve the capacity to assess, evaluate and select technologies appropriate to local conditions and the development needs of the country as a whole. The Bank will encourage research and the use of pilot or experimental programs to develop and test innovative technologies.

International Cooperation

7. The Bank shall cooperate with regional and international agencies to promote the development and dissemination of intermediate technologies. In this regard, the Bank shall help to fund seminars, sponsor training programs and pilot projects, participate in interagency consultations, foster the exchange of information, and promote the establishment of institutional arrangements in Latin America which would be responsible for the development of intermediate or light-capital technologies and the dissemination of information to member countries.

The Brazilian Gasohol Programme

by

Victor YANG & Sergio C. TRINDADE, Brazil

ABSTRACT

The striking dichotomy in Brazil's energy picture is her simultaneous heavy dependence on imported energy and a historical reliance on hydropower and biomass.

Gasohols — gasoline and alcohol (ethanol) fuel blends — have been used in auto vehicles in Brazil since the 1930's. Compulsory addition of alcohol to gasoline started in 1931 to hedge the important sugar industry against sugar price fluctuations in the international market.

Many years of experience and testing have confirmed that all Otto engines manufactured in Brazil maintain a desirable performance when fueled with gasohols containing up to 20% alcohol compared with operation on straight gasoline. Some vehicle fleets operate today successfully with straight alcohol, but the engines require minor changes to take advantage of the higher compression ratio (12:1) allowed by straight alcohol.

A national alcohol program was launched by the Government in 1975 aimed at increasing alcohol production for the fuel and chemical feedstocks markets. A tentative 20% alcohol level in gasohol, uniform all over the country, is a goal already achieved in some areas such as the City of São Paulo (1.3 million auto vehicles).

Although not explicit in the original decree, the Brazilian Gasohol Program may help achieve important social, economical and political objectives.

As of November 1978, 208 new distillery or distillery expansion projects have been approved adding 3.9 million m^3/year of alcohol production capacity within 2-4 years. Sugar-cane is the primary feedstock for fermentation alcohol production. However, sweet sorghum and starch containing materials such as mandioca (cassava, manioc), and babassu (the fruit of a Brazilian native palm tree) will rapidly grow in importance as prime land becomes scarcer.

Process technologies will require considerable improvement as alcohol production expands, and its primary market changes from cosmetics and pharmaceuticals to fuel.

Process energetics measured by net energy ratios indicate for the best alcohol producing system configuration a high return on energy investment (NER around 8:1).

Process economics estimates indicate a cost of $ 360-380/m^3 FOB distillery ($ 1.36-1.44/gal) against an administered alcohol price close to $310/$m^3$ ($1.17/gal). This price differential constitutes the basis for Government subsidies. Agricultural

144

and industrial conversion improvements will be required to remove the need for subsidies in the medium term. Gasoline at the gas station sells for $390/m^3$ ($1.54/gal).

The environmental consequences of producing fermentation alcohol may be serious. Stillage (bottom slops) is produced at a rate of $13m^3/m^3$ of alcohol and is highly pollutant. There are several stillage processing alternatives, which could turn a problem into an opportunity. There are also many opportunities for new concepts in stillage treatment and separation of alcohol from fermented worts.

A temporarily depressed world sugar market and market guarantee for fuel alcohol have favoured conversion of sugar-cane to alcohol in increasing quantities. Nevertheless, this short term optimism in the achievement of the Brazilian Gasohol Program can only be sustained if timely concerned efforts by Government and Industry are set forth to overcome the challenges ahead.

Major issues such as limited availability of subsidized financing, marginal economics, high investment requirements, high production costs, agricultural and industrial management problems, difficulties with storage and distribution of alcohol, etc. will be key to the pace of the Brazilian Gasohol Program hereon.

Only those countries with sufficient land mass relative to demand, sunshine and water resources to grow food and feed at a surplus level, could probably resort to gasohol programs to help alleviate energy problems. Such is the case of Brazil and the United States of America.

1. INTRODUCTION

The striking dichotomy in Brazil's energy picture is her growing dependence on imported energy coupled with a historical reliance on hydropower and biomass.

Brazil's accelerated industrial development during the last two decades has been accompanied by an increasing demand for imported energy, mainly as petroleum. In 1977, domestic energy sources accounted anly for 60% of the country's primary energy consumption.

A lagging domestic petroleum production coupled with a heavy dependence on liquid fossil fuels for transportation have led to a petroleum import bill of almost $4 billion in 1977. This tab alone represented 31% of Brazil's total import expenditure and surpassed export revenues generated by traditional items such as coffee and sugar. Since the petroleum price hike in 1973, Brazil has faced problems with balance of payments and a slowdown in the country's GNP growth rate.

Petroleum accounts presently for roughly 43% of Brazil's primary energy consumption. Domestic production in 1977 supplied only 16% of the country's petroleum needs. In a period of roughly a decade, petroleum has displaced biomass as the major primary energy source. Nevertheless, hydropower and biomass are important energy sources in Brazil even today. In 1977, roughly 25% of total primary energy consumption was in the form of hydroelectricity whereas biomass accounted for about 28%. Wood, sugar-cane bagasse, charcoal and fermentation alcohol (ethanol) are the traditional biomass fuels in Brazil.

2. THE BRAZILIAN EXPERIENCE WITH ALCOHOL AS FUEL

Gasohols — gasoline and alcohol fuel blends — have been used in automobiles dor decades in Brazil. The earliest recorded experiments with alcohols and gasohols date back to the '20s and '30s. Compulsory alcohol addition to gasoline to a level of 5% was legislated in 1931 motivated mainly by the Government's interest in stabilizing the sugar industry, one of the foremost sectors of the Brazilian economy. Excess molasses and sugar were converted to alcohol in distilleries attached to sugar mills. The alcohol content in the gasohols, however, oscillated considerably depending on the domestic and international sugar and molasses market.

Pioneering experience with straight ethanol (96° GL) fuels in automobiles were performed by INT (National Technology Institute) in the 1920's. Between 1974 and 1976, CTA — the Air Force Technical Center, investigated the performance of Brazilian engines fueled with gasohols containing varying amounts of alcohol. Both bench and road tests were carried out. The overall conclusion was that all existing Brazilian Otto-engine vehicles, when fueled by gasohol containing up to 20% (vol.) of alcohol, did not lose performance compared with straight gasoline operation. Furthermore, no engine modifications were required.

Most Brazilian engines operate with a fuel rich mixture (fuel/air ratio above stoichiometric) which help to explain why driveability and fuel-economy have not been greatly affected by gasohols containing up to 20% alcohol. The presence of alcohol has raised the octane rating of the average Brazilian gasoline. For example, Brazilian gasohol with a 20% alcohol content shows typically an octane rating of 81 MON, up from 73 MON for the base regular gasoline.

Climatic conditions do not hamper the performance of gasohol fueled engines in Brazil. Engine start-up and phase separation problems have not been reported mainly because ambient temperatures are seldom below freezing.

Starting in 1977 the City of São Paulo (approximately 1.3 million Otto-engine vehicles) and other large urban centers were fueled with gasohol containing close to 20% alcohol. This pattern was apparently extended to the entire State of São Paulo beginning in 1978.

Otto-engine vehicles fueled by straight alcohol (96° GL ethanol) are also being tested in Brazil. TELESP's (São Paulo Telephone Company) fleet of 400 VW with modified engines has been operating in the City of São Paulo since 1977. These vehicles operate with a higher compression ratio (12:1) than the standard practice (7:1). Maintenance vehicles belonging to telephone, electrical utility companies and other Government institutions, totalling 725 vehicles, were operating with straight alcohol up to October 1978.

Gasohol preparation is carried out in one of two ways. Some distributors have blending tanks whereby gasoline and alcohol are mixed prior to truck loading. Other distributors do not have a separate blending tank and addition of gasoline and alcohol is flow controlled during the truck-filling operation.

Currently, there are tests underway in Brazil to utilize alcohol also in Diesel-engines. If these tests are successful, fermentation alcohol can play an even larger role in the Brazilian energy market. The potential represented by Diesel oil substitution with alcohol may be appreciated from the fact that diesel oil and gasoline are consumed at similaar leveeeels in Brazil.

3. BACKGROUND AND OBJECTIVES

The turn of events in 1973 prompted the Brazilian government to foster the development of domestic energy sources. In this context, the Brazilian National Alcohol Program — hereafter referred to as the Brazilian Gasohol Program — was established in November of 1975. The primary stated objective is a rapid increase in the supply of fermentation alcohol from all sources for usage as fuel and as chemical feedstock.

Alcohol production, historically linked to the sugar industry, should gradually develop into an independent venture in many locations, based on both sugar and starch-containing feedstocks.

An increasing number of independent distilleries (those not connected with production of sugar or any other commodity) are therefore expected. In addition to molasses, sugar-cane juice and mandioca (cassava, manioc, better known in the U.S. in a granular from called tapioca) are alternative feedstocks for Brazilian distilleries. Babassu (a native coconut) and sweet sorghum may also become alcohol feedstocks in the future.

The departure from the traditional byproduct status for alcohol to one of major product should bring about important economic, social, political and technological repercussions in Brazil. Improvement of both industrial and agricultural technologies will require intense R & D and managerial efforts, considering alcohol production in unprecedented scales. Alcohol production from mandioca and other non-traditional fermentation sources may lead to agro-industrial activities in remote areas. The energetics and economics of alcohol production are also noticeably different in the case of direct alcohol production in the independent distilleries.

Important social benefits expected from the Program include creation of jobs at the farm level and thereby a possible reversal of the urban migration trend in the long run, improvement in income distribution and a more balanced development throughout the country.

Economic objectives being sought with the Program include savings in scare foreign exchange through substitution of imported fossil fuel and strengthening of the internal market through an intensified usage of domestic production factors. For example the Brazilian industry is supplying most of the equipment needs for the new distilleries.

Political implications of the Program include the occupation of idle and under-utilized territories and the long term prospect of relative independence from external sources of energy and chemical feedstocks.

4. STATUS OF THE PROGRAM

Substantial increase in fermentation alcohol production has been achieved in Brazil since the implementation of the gasohol program. Low international sugar prices coupled with government incentives enabled an alcohol production of close to 1.5

million cubic meters in 1977, a 100% increase in a year. Nevertheless, this production increase resulted mainly from utilization of idle capacity of existing distilleries (installed prior to 1975) and allocation of substantial amounts of excess sugar for alcohol production.

Up to November 1978, 208 distillery projects have been approved by the National Alcohol Committee (CNAL) totalling some 3.9 million m^3/year of additional capacity and requiring over $1 billion in terms of industrial investment alone. Of the capacity to be added, roughly 33% represent expansion of existing units and 67% are accounted for by grass-roots units. This additional capacity should be in operation within 2-4 years.

In terms of geographical distribution, the total nominal capacity in the Northern/Northeastern regions should increase by approximately 1.2 million m^3/year or 31% of Brazil's total. The authorized additional capacity for the Central/Southern regions amount to 2.7 million m^3/year or 69% of the country's total. The State of São Paulo, traditionally the largest sugar and alcohol producer, has an authorized additional capacity of approximately 1.5 million m^3/year.

Starting in 1977-1978 sugar-cane season, 55 new distillery projects (including expansion of existing units) were scheduled to come on stream, representing a combined capacity of close to 1020 thousand m^3/year, in addition to the total installed alcohol distillery capacity of close to 1 million m^3/year, existing prior to the institution of the Brazilian gasohol program. The new distilleries being erected have a capacity in the range of 30 - 650 m^3/day. Average size falls close to 120 m^3/day.

In terms of carbohydrate feedstock, molasses or sugar mill distilleries account for 47% of the total additional capacity authorized as of November, 1978. Independent distilleries such as sugar-cane juice distilleries account for 44% and mandioca distilleries for 9%. To-date one babassu distillery with a capacity of 9 thousand m^3/year has been approved. Mandioca distilleries are to be located mainly in remote areas, near mandioca farms which can be grown under soil and climatic conditions often improper for sugar-cane.

The first mandioca distillery with a capacity of 60 m^3/day was erected by Petrobrás with know-how developed by INT (National Technology Institute) for demonstration purposes, and is presently undergoing start-up tests.

5. THE BRAZILIAN ALCOHOL MARKET — AN OVERVIEW

By 1985, total fermentation alcohol production in Brazil is expected to be in the range of 4-6 million cubic meters. If all the alcohol thus produced were to be used in fuel blends, it would supply approximately 2% of Brazil's primary energy requirements in 1985, according to projections by the Ministry of Mines and Energy.

Both mandioca and sugar-cane yield roughly 3.3 m^3/ha/year of alcohol. Consequently, a production of 5 million m^3/year by 1985 would require at least 1.5 million hectares. This area is smaller than the currently cultivated area for sugar production in Brazil and represent less than 0.2% of the surface of the country. In the short term, most of the fermentation alcohol production is going to the gasoline pool. Although not stated in the original decree, the Brazilian Government appears to be striving for an alcohol production adequate for a gasohol with 20% alcohol content, on a country-wide basis. This target appears feasible by the mid 80's, if the pace of the gasohol program is maintained and assuming conservative growth rate for alcohol consumption in the chemical industry.

Industrial alcohol not going to the fuel market finds application as solvent and cleansing agent in the pharmaceutical and cosmetics industry and as feedstock for the chemical industry. Consumption of alcohol as a chemical feedstock is expected to outgrow considerably the quantities used by the pharmaceutical and cosmetics industries by 1985.

The existence of a guaranteed market for fuel alcohol in the near term constitute a major incentive for the implementation of new distilleries and greatly influence the location of these units.

6. PROCESS CONSIDERATIONS

6.1. Process Technology

Batch fermentation processes are currently employed by the Brazilian distilleries. Alcohol production from molasses is similar to the sugar-cane juice-derived alcohol, with a molasses dilution step prior to fermentation.

Alcohol production directly from sugar-cane entails a juice preparation step which is carried out mechanically through sugar mills, prior to fermentation and followed by distillation. Large quantities of bagasse and stillage are produced as by-products. A small quantity of fusel oil is also generated.

Alcohol production from mandioca entails preparation stages to convert the starch present in the roots into fermentable sugars. After the initial washing and size-reduction steps, the starchy mash is liquefied (converted to dextrines) through a combination of thermal (steam cooking) and enzymatic actions. α-amylase is added prior to the cooking step. The liquefied mash is in turn hydrolized to fermentable sugars in the saccharification step through the action of the enzyme glucoamylase. The fermentation and distillation steps are similar to the sugar-cane-derived alcohol. Stillage is the majorby-product with fusel oil, mandioca skin and fibers generated in lesser amounts.

Presently, most of the sugar-cane-based distilleries burn their bagasse as boiler fuel for steam raising and electricity generation and the fusel oil is sold.

Large energy consumption is incurred specially in the distillation step. In sugar-cane-based distilleries, bagasse availability is adequate to provide for energy self-sufficiency. The situation in mandioca distilleries is different due to the absence of "bagasse" if only the starchy roots are processed. However, if total mandioca harvesting is practiced the stalks may be a source of energy.

149

6.2. Process Economics

The mandioca distillery turns out to require a fixed investment which is 20% higher than an equivalent independent sugar-cane distillery. Extra equipment for liquefaction and saccharification accounts for the difference.

Larger alcohol inventory for the independent sugar-cane distillery due to fuel alcohol legislation entails a considerably higher working capital for the sugar-cane distillery. As a result, total investment differ by less than 10%.

The alcohol selling prices were computed including a return on investment of 12%/year, DCF method. Economics of mandioca alcohol and sugarcane alcohol on the basis of alcohol selling prices differ by less than 10%. The administered price of fuel alcohol currently close to $310/m³ ($1.17/gal) is lower than the calculated selling prices for mandioca and sugarcane alcohol, in the range of $360 - $380/m³ FOB distillery ($1.36 - 1.44/gal). However, if subsidized financing available under the Program is considered, fermentation alcohol economics appears marginally attractive.

Breakdown of the alcohol selling prices shows a striking economic similarity. Agricultural feedstock accounts for the larger fraction of the estimated selling prices. Hence, high sensitivities of alcohol economics to agricultural and distillery yield increases are expected.

By-product credit in both cases appears small. Note, however, that the raw stillage was valued mainly as a mineral fertilizer substitute. If the proper value of the organics in the stillage is assigned or other more efficient stillage upgrading methods are considered, it may be possible to generate a higher net by-product credit. Moreover, excess bagasse in independent distilleries, available above steam raising needs, may be a source of additional by-product revenues if proper pricing and market conditions are found.

6.3. Process energetics

The net energy ratio (NER) analysis relates energy contents of output streams to that of input streams. Proper energy accounting includes all energy consumption associated with input streams such as fossil fuels associated with fertilizers used in the sugar-cane or mandioca plantations.

Solar energy is obviously the major energy input, but it does not enter NER calculations because it is freely available.

The system configuration for NER analysis of alcohol production therefore includes both agricultural and industrial subsystems. The energy-containing streams crossing the boundaries of the system are topically electricity, fuels, fertilizers, chemicals, agricultural products and human labor. The most important ones, considering the mandioca industrial subsystems, are firewood and/or stalks (for steam raising) and electricity.

For a typical 150 m³/day alcohol distillery, NER for the best system configuration reaches 4-5 for both mandioca and sugar-cane alcohol.

6.4. Process ecologics

Stillage (bottom slops) is the largest volume by-product of alcohol distilleries which comes off the bottom of the first distillation column. It is generated at a rate which is 12 to 13 times in volume that of the alcohol production. It has been estimated that total stillage output by the Brazilian distilleries in 1977 was in the order of 20 million m^3.

Because of its high content of soluble organics and inorganics, stillage has a high pollution potential if discharged into rivers. For example, the total Brazilian stillage output in 1977 was roughly equivalent to the sewage corresponding to 50 million inhabitants in term of BOD.

Since stillage normally does not contain pathogenic bacterias or viruses, nor heavy metals or polychlorinated organics, recovery of its minerals and organics is a potentially attractive undertaking. It is technically feasible to convert stillage into marketable products such as fertilizers and feed additive or into methane as a supplementary energy source. Economics will ultimately determine the choice of stillage treatment process and the product form.

Presently there are two stillage treatment processes being used in Brazil. The most widespread practice is lagooning. The other process takes advantage of the fertilizer value of stillage. Raw stillage is sprayed over the sugar-cane fields, thereby reducing or eliminating the mineral fertilizer requirements.

Larger-size and increased number of independent distilleries call for more efficient and economic means of stillage utilization. Some of these non-conventional processes being proposed and assessed are:

- stillage concentration through either multiple effect evaporation or mechanical vapor recompression for the feed market or as fertilizer (up to 60% total solids);
- incineration of concentrated stillage for ash recovery;
- anaerobic fermentation for methane generation;
- aerobic fermentation for high test single cell protein (SCP) production.

Advanced separation technology such as membrane based processes are also being investigated at the laboratory level.

Proper treatment of stillage may improve the alcohol distillery's economics, if adequate price levels and markets can be found for the stillage products.

7. PROBLEMS AND OPPORTUNITIES

The Brazilian Gasohol Program faces a number of challenges in the near to long terms.

Uncertainties of various kinds are plaguing the program such as:
- lack of clear political decision and commitment to carry the program to its fullest extent;

- limited availability of subsidized financing;
- lack of clear commitment of the Brazilian auto industry to gasohol, straight alcohol, and to the use of alcohol in Diesel engines;
- effect of gasohol on oil refining schemes, generating surplus gasoline that has to be exported;
- lower ROI of alcohol compared to gasoline;
- alcohol storage and distribution problems;
- expansion of existing distilleries (thereby favouring concentration) versus grassroots distilleries in unexplored regions (favouring interior development and hence decentralization);
- high investment, high costs and marginal economics of fuel alcohol production;
- development of proper agricultural and industrial technologies as alcohol evolves from a sugar by-product to an energy product for the fuel market;
- lack of a firm coordinated energy policy at a national level, encompassing both fossil fuels and renewable resources.

Alternative and apparently cheaper short term energy strategies are frequently cited by critics of the program. For example, fuel oil conservation programs could save Brasil in 2 years as much petroleum with a much smaller investment than would the gasohol program in 7 years.

The economics of alcohol production in independent distilleries appears to be unfavourable. Reduction of investment and operating costs are therfore essential if the Program is to attain the scale proposed.

Movement of large volumes of agricultural feedstocks, stillage and alcohol also pose considerable technical and managerial problems. Proper siting and coordination of the agricultural and industrial operations are important for the ultimate succes of distillery ventures.

Brazil's agricultural base will need strengthening, if alcohol feedstocks are to achieve more acceptable quality standards and cost levels. Plant genetic research and optimization of farming practices appear to demand considerable R & D effort in the upcoming years.

Improvement in the distillery's process and operations' technology may be critical to the feasibility of alcohol production. Large scales of operation tend to favour continuous processing. For example, continuous liquefaction and fermentation warrant concentrated R & D efforts. Energy-saving separation processes such as reverse-osmosis may eventually substitute conventional distillation and evaporation operations.

The State of São Paulo alone accounted for close to 2/3 of Brazil's total alcohol production in 1977. If other states do not keep up with the alcohol producing pace of São Paulo, an even more serious polarization effect may result. This scenario would go against the policy of industrial descentralization, one of the underlying objectives of the Brazilian Gasohol Program. In the near term, therefore, alcohol distribution is one of the largest challenges that Brazil faces. Starting in 1978, excess fermentation alcohol in São Paulo after allowing for state-wide usage of gasohol with 20% alcohol were trucked to neighbouring states.

Local alcohol supply/demand unbalance problems in the fuel market coupled with local shortage and escalating prices of petroleum derivatives may lead to increased usage of alcohol in the Brazilian chemical industry as feedstock.

Ethylene, acetic acid and octanol are currently manufactured from alcohol. A number of new units are being planned.

8. CONCLUSIONS

The scope of the Brazilian Gasohol Program is bound to have profound implications for the country in the long run.

If successful, the Program has the potential of providing Brazil with a higher degree of energy and technology independence, in addition to economic, social and political improvements.

Agriculture may be the basis for Brazil's needs not only for food, fiber and feed, but also for chemicals and energy due to a favourable combination of land area, latitude, population and stage of development.

The substantial increase in alcohol production achieved since the implementation of the Brazilian Gasohol Program appears to indicate that large-scale substitution of fossil fuels may be achieved and sustained in the near future.

Nevertheless, necessary improvements in alcohol economics, agricultural and industrial technologies, means for storage, and distribution of alcohol challenge Brazil's technical and managerial skills.

Economics of direct alcohol production in independent sugar-cane and mandioca distilleries appears to be marginal at present, even accounting for Government subsidies. It is not surprising therefore, that most of the fermentation alcohol production today derives from sugar mill distilleries with alcohol playing a by-product role. Moreover, a depressed international sugar market has favoured conversion of sugar and molasses to alcohol and further contributed to the high production levels already recorded.

Alcohol production from sugar mill distilleries cannot, however, attain the quantities required for fuel and chemical usage at a national level. Moreover, difficult-to-predict sugar market oscillations and distribution problems may be critical. Already the State of São Paulo, the largest sugar producing state, faces a surplus of alcohol which needs to be trucked to other states.

Sustained large-scale alcohol production, therefore, can only be achieved if a large number of independent distilleries, based on locally available feedstocks, are erected throughout the country. In this context, improvement in agricultural and industrial technologies for conversion of sugar and starch-based alcohol feedstocks are key to the economic feasibility and long-term success of the Brazilian Gasohol Program.

Large-scale alcohol usage as chemical feedstock may alleviate some of the regional fuel alcohol market imbalances and help consolidate the Brazilian Gasohol Program.

Gasohol programs should not be generally recommended to all countries in the world. Probably only those countries with sufficient agricultural land, sunshine and water resources to grow food, feed and fiber at a surplus level could resort to gasohol programs to help alleviate energy problems. Such is the case of Brazil and the United States of America.

Solar Energy and Desalination

by

Dr. Javier Ibarra HERRERA, Mexico
Director General for the Use of Saline Water and Solar Energy

Water and Energy Development go hand in hand making possible the establishment of industries in human settlements, permiting local growth and wealth, promoting jobs and restraining migration in search of better opportunities.

Where water availability is scarce or has reached its exploitation limits, or in areas where it is available but is saline or improper for use, development becomes a very difficult problem as well as an expensive matter to be solved. For many years, transportation of water by acueducts has been the most used solution, causing sometimes the reduction of the growth potentials of other areas, provoking a unabalance that results in large cities, like Mexico City, for which the increasing demands become great problems of social, economic, technical and political nature.

My country, Mexico, has 10 000 km of coasts and a marked imbalance between human settlements and water availability, 67% of the 2 000 000 km² are arid, semiarid or desertic, and 8% of the population has 40% of the hydraulic resource, while 60% has only 12%.

Large areas, and coastal locations have a limited growth because of the lack of energy and water, though richness of the sea and land, can be exploited profitably, with this essential elements.

An office for the desalination of water was opened in 1971. This technical and administrative organism is meant to meet and solve problems related with the use of saline waters, and since it had the technical background, planning experience in remote areas, operation, maintenance and design capabilities which were similar to the needs of a solar energy department, the government appointed this duties to the "Dirección General de Aprovechamiento de Aguas Salinas y Energía Solar", of which I am General Director.

We have 45 units of different types, which are Reverse Osmosis for brackish and sea water, multistage Flash, Vapor compression and solar distillation. We also have 17 solar pumps.

And one low temperature electric plant of 25 Kwe that is used to pump water.

Within the activities of the General Direction we have negotiated binational agreements of which the one with Germany comprises:

A) Project of a Solar Multi Stage Flash.

B) Project of a Solar Brackish Reverse Osmosis System.

C) The Project Sonntlan that is a solar integral system to provide a town of 250 people with, electricity, desalinated water, ice, refrigeration, solar houses, etc.

As for solar desalination, the General Direction has 3 installations which we call "fields" to produce 1 to 1.5 m³/d of fresh water from sea water. They are located between the paralels 24 and 30 in the Gulf of California and the Pacific Ocean. This fields are made of the typical distiller with glass roof, fiber glass body and produce an average of about 5 lts/day/m². Although of simple design and apparent self-sufficiency, we have experienced difficulties with this installations, primarily because of idiosyncrasy, the installation belongs to all the people but at the same time to no one in particular. Cost wise the production of fresh water is also high, due to diffcent factors depending on the locality and the availability of the necessary materials for good operation.

Due to this problems, we conceived the family type distiller for minimum water requirements, that is a train of distillers equal to the number of members in a family. With this, the head of the family becomes involved as the responsible for the installation and the autosupply of his family water needs. A pump is not required, a small deposit is filled up using buckets autoconstruction techniques are applied, and care for fragile parts is more taken into account.

Nevertheless the General Direction is not satisfied with this results and has been in the process of studying other alternatives.

So in 1977 in a visit to Germany, the possibility of uniting efforts in a project for a solar powered multistage flashunit, idea that had been studied by us and proved attractive, was then cristalyzed in an agreement. On the German part, Dornier System was chosen as the partner, a firm which has been involved in solar desalination since 1972.

The design of the MSF was done in Mexico, while the solar boiler in Germany. The unit will be built in Germany shipped to Mexico, to the city of La Paz, where it will be tested, operated and evalvated.

The German Government through the Bundes Ministerium für Technologie (BMFT) and in particular Dr. Helmut Klein, head of the Non-Nuclear Energy Research and Technology Section, are a very enthusiastic group, that understanding new ways of technical cooperation, have supported this interchange on an equal basis and for mutual benefit. It is my privilege to pronounce our great satisfaction in this cooperation and the excellent relations with our German partners.

The solar energy MSF-pilot plant

Design

A pilot plant was specially designed for a large range research program to investigate the operation characteristics of the entire system under different conditions of solar radiation, sea water analysis, sea water and process top temperatures, various part loads, etc. The plant has a capacity of approx. 420 l/h at a sea water max. temperature of 100 °C and the specific heat consumption is approx. 95 kcal/kg.

The plant is designed for a maximum sea water process temperature range between 80 and 100 °C. The sea water flow is recirculated.

The evaporator consists of 10 identical stages, the shell material being cunifer.

156

The collector area is approx. 300 m^2 which is sufficient to supply the desalination plant with the energy necessary for 24 hours per day of full load operation. Hot water tanks to store the energy for operation at night are part of the system.

Process description

The pilot plant operates according to the multi-stage flash evaporation (with brine recycle) principle.

MSF - plant

Sea water is heated in a heat exchanger (brine heater) by means of hot water of the solar system and then flows to the flash chamber of the first evaporator stage. A part of the super-heated sea water is flashed off and the resulting vapour is condensed on the outside of the condenser tubes, which carry the lower temperature sea water to the brine heater. The remaining part of the sea water flows into the subsequent stage, where further flashing occurs and further distillate will be produced. This process is repeated in all stages of the evaporator. In the last stage the lowest temperature and vacuum is achieved. To obtain and maintain the vacuum in the evaporator and to pull out the noncondensable gases of the evaporator, a vacuum pump is provided. The concentrated brine and the distillate are discharged by pumps from the last stage.

Solar system

The hot water is recycled the whole day through the collectors from hot water tank 2 to hot water tank 1 and will be heated by the solar energy. A part of this hot water will be stored in tank 1 for the night hours and the other part flows to the brine heater to heat the seawater to its top temperature. During the night the stored hot water flows from tank 1 through the brine heater to tank 2 and heats the sea water to its top temperature. This hot water circuit ensures a continuous operation of the MSF-plant for 24 hours per day at constant load.

Operation characteristics of the pilot plant

When the collected solar energy is constant and hot water outlet temperature of the collectors is constant, the MSF-plant will be operated with constant load. In case of bad weather, when the input of solar energy is low and respectively the hot water outlet temperature of the collectors is low, the sea water top temperature decreases accordingly and the production of the MSF will be reduced automatically. For a long period of bad weather, where the collected solar energy is too low, the MSF plant must be shut down or the auxiliary oil fired steam boiler must be used to keep the MSF-plant on stream.

Specifications

Productivity	330-420	l/h
Brine maximum temperature	80-100	°C
Specific heat consumption	95	Kcal/kg
Purity of distillate	less than 40	ppm
Electrical energy for pumps	5	KW aprox.

Another means of using solar energy for desalination, is to cover the electrical demand of an R. O. system, either to desalinate brackish or sea water.

With GKSS and AEG in Germany appointed by the BMFT, we are also developing a brackish solar reverse osmosis unit that we call SORO. This unit will be tested in our regional unit in Concepción del Oro, Zacatecas, areas in which we have great amounts of underground brackish water, in the semidesertic and arid zones of Coahuila and Chihuahua.

This is a small plant to fulfill primary necessities which power is supplied by fotocells. Again German and Mexican technicians are working to minimize the energy requirements, as well as to make a system easily maintainable.

Both units will start to operate in the later part of this year.

We are talking with the French Government to develop a series of cooperation programs, among which we are studying the possibilities to desalinate sea water with solar energy.

The solutions to the problems in desalination and solar energy have just begun. The way is long and the challenge is great, but we have a firm conviction of the future that stands before us, and that generations to come, will find that the efforts of today will be a key to their needs.

"Helio-Technologies in Perspective: A Brief Guide to their Utilisation"

by

B.A. GIBBS, Guyana

The present position is that there are a host of helio-technologies with few guidelines for their successful implementation. It appears that much of the promise the new helio-technologies hold is quickly fading in the eyes of their users because of rash and inappropriate implementation and because of inflated expectations created by world-wide publicity they have been receiving.

Technologies should be recommended on the basis of the state of the art, not on their imagined impact. The fact is that for more reasons than just lack of expertise many of the technologies are too underdeveloped to be applied in developing situations. Much of the so called appropriateness of the technologies disappear when the requirements of scale are examined. Therefore, for all applications it is useful to examine the requirements of the technological route in an attempt to evaluate the use of one route against another for a given application. Only the following conversions will be examined viz. the production of

    (a) Electricity
    (b) Mechanical Energy
    (c) Thermal energy (heating/cooling)

starting with

    (i) sunlight
    (ii) wind
    (iii) water
    (iv) biomass

and (d) a few other selected conversions.

The objective is to compose a quick reference to the main features and advantages/disadvantages of a particular route.

Energy Production starting with Sunlight

The simplest technical route for electricity production, is from sunlight utilising the photovoltaic device or solar cell. At present this is still however a highly expensive route per kilowatt of electricity produced.

Nevertheless in terms of maintenance and manpower requirements it is the most trouble free route, everything depending on the solar insolation and efficiency of the unit when matched to the system it is to drive. This method of electricity production has been found suitable for remote meteorological stations, buoys, rural educational T.V., water pumps and many other applications where a few kw are required.

159

Thermal routes require the matching of a solar thermal generator to a Rankine-cycle engine or external combustion engine such as a free piston engine. This is a more involved route and may require, depending on the scale of operation, a special working fluid in a closed cycle. Normally the designer has to depend on a steam engine/Rankine-cycle engine since the technology is easily available compared to the stirling or free piston engine types. Matching of a flat plate collector system using water, to a turbine/compressor cycle may be solved by utilisation of an organic working fluid in a closed turbine/compressor cycle.

The best utilisation of solar flat plate collector systems is in the production of hot water or air for space heating and drying applications.

The production of mechanical energy from sunshine is also cumbersome although this route is used with varying success for solar water pumping.

Energy Production starting from Wind

Only the production of mechanical energy and electricity are considered. Both of these are highly recommended routes depending on the application. For mechanical power the load should be able to operate under variable speed conditions. For electrical energy the route is considered suitable only for small scale application and in energy systems where it can contribute to the saving of fossil fuel or for mixed helio-systems.

Energy Production starting from Water

Only the production of mechanical energy and electricity are considered. Given the hydro-requirement use of the water wheel, because of its ease of fabrication at the village level, seems to be the best means of meeting small to medium scale requirements for mechanical energy for milling. Small turbines are also the easiest means of acquiring electricity given the hydro-requirement.

Hydro-power requiring dams are also highly recommended.

Energy Production starting with Biomass

Normally biomass systems are desirable from the point of view of waste disposal. There are three methods of using biomass viz:

1. for simple burning
2. for the production of producer gas or methane
3. in bio systems that produce organic fuel or plants in the presence of sunlight.

Burning is not considered as it is simple and much practised, although enough use is not made of new methods such as fluidised combustion techniques, waste heat utilisation etc. Pyrolysis is considered under 'other technologies'.

160

Gas producers fall in the class of known but presently still being developed, technologies. One commercial range of wood burning gasifiers is known to be produced by a German firm. The use of gasifier units is highly recommended given the wood waste. The development of such gasifiers should allow utilisation of a wider range of organic waste.

The technology surrounding biogas generators is widely applied in China and India although it is much reported that the biochemistry of digestion is not fully understood. Experience in utilisation of such units needs to be transferred in the case of most applications. Having generated gas from biomass wastes the simplest uses are for space heating, cooking and illumination. However gas engines may be used to produce mechanical power. This may be done by conversion of standard diesel units for the production of mechanical power for water pumping and village industry, or for the production of electricity.

Solar Air-Conditioning System
for the Barbados Central Agronomic Research Unit

by

B. McNELIS, U.K.

ABSTRACT

This paper presents a summary of a recently completed study of the application of solar air-conditioning to the Central Agronomic Research Unit (CARU) which is to be constructed for the Barbados Ministry of Agriculture. The work has been performed for the Barbades Government and financed by the European Development Fund (EDF). A number of systems using solar and wind energy, and a combination of both, were examined but none of these could be justified on economic grounds. However, when taking other non-economic considerations into account a system using photovoltaic cells has been recommended.

INTRODUCTION

In 1977 a decision was taken by the Barbados Government to seriously consider the incorporation of solar air-conditioning systems into three new buildings planned for the Ministry of Agriculture. In addition to CARU the other buildings are the Administration Building which is to be located alongside CARU at Graeme Hall and the Laboratory of the Government Analyst which has a site adjacent to the Barbados National Standards Institute at Flodden. The building and equipping of CARU is to be financed by EDF and construction is expected to commence later this year. Construction of the other two buildings is already well underway and these will both incorporate solar air-conditioning systems employing lithium bromide/water absorption chillers and flat plate solar collectors supplied by Tadiran of Israel.

The study included an analysis of the outline designs for CARU which had been prepared by the architects, Robertson Ward Associates, leading to advice on how to reduce the air-conditioning load by passive means, followed by a computation of the load. A number of options for the air-conditioning system, using conventional energy and solar and wind energy were identified and compared technical and economic viewpoints.

BUILDING DESIGN AND COOLING LOAD

The building has a floor area of about 2 000 m^2 on two storeys arranged around a central courtyard. The energy requirements for air-conditioning have been minimised by:

- providing for part of the building on the windward side to be naturally ventilated and so not require air-conditioning
- use of naturally ventilated roof spaces

162

- the use of a 1.5 m overhang on the north and south facades to reduce solar gain
- the use of 1.2 m overhangs and "egg crate" shading on east and west facades.

The cooling load was calculated on an hour-by-hour basis (by computer) using weather data recorded by the Caribbean Meteorological Institute in Barbados. The peak cooling rate required was found to be 77 kW and typical daily cooling loads varied between 212 kWh in February and 274 kWh in September, the total annual load being 3 063 kWh.

ENERGY AVAILABILITY

No insolation or wind measurements have been made at the site, and data from Grantley Adams airport has been used.

Total global radiation (horizontal surface) varies between 5.0 kWh m^{-2} in November and 6.4 kWh m^{-2} in April with an annual total of 2 082 kWh m^{-2}. No measurements of the diffuse and direct component have been made but using the Liu and Jordan method diffuse radiation was estimated as between 27% in December and 34% in August. It is considered that the use of data not obtained at the site is acceptable for a system design.

Because the output of an aerogenerator varies with the cube of the wind velocity, which itself can vary considerably from place to place, the energy delivered by an aerogenerator is difficult to predict. Thus it is not possible to make a definitive comparison between wind and solar based systems although a wind system outline was prepared. The wind climate in Barbados is characterised by a higher average wind velocity between January and July than between August and December.

The data used, from Grantley Adams airport, indicated an average wind velocity which varied between 3.75 m s^{-1} in October and 6.58 m s^{-1} in March and June, corresponding to intercepted energy of 33 W m^{-2} and 280 W m^{-2} respectively.

AIR CONDITIONING SYSTEM CONSIDERATIONS

A range of options for the air-conditioning system were considered. These were:

(i) Electrically-driven Vapour Compression Chiller
(ii) Gas-driven Absorption Chiller
(iii) Solar/Absorption Chiller
(iv) Solar/Rankine Cycle Vapour Compression Chiller
(v) Photovoltaic Cells/Vapour Compression Chiller
(vi) Solar-Sorbent/Evaporative Cooling
(vii) Wind/Vapour Compression Chiller
(viii) Solar + Wind/Absorption Chiller.

The two conventional systems are (i) and (ii) and in the absence of a policy to employ a solar derived system the obvious choice would be the electrically driven vapour compression chiller. The absorption chiller would not normally be used as this involves both higher capital and running costs.

163

The non-conventional options were compared with respect to availability of components, performance and cost. Commercially available components were not identified for the evaporative cooling with dehumidification or Rankine engine approaches and so these were not examined further. However, it was felt that these approaches, particularly the latter, would be important in the future.

Because both wind and solar energy availability have their minimum at the same time of year, which is also the time of peak cooling load, the combined wind and solar system offers no advantages over a wholly wind or solar system.

Hence the ambient energy which were seriously considered were options (iii), (v) and (vii).

Solar/Absorption Cycle

The absorption cycle for cooling uses heat as the energy input and therefore there is the obvious possibility for the use of solar heat as the energy source. Because absorption chillers are commercially available, this approach has received by far the most attention. Solar collectors of many designs are now in an advanced state of development and are also commercially available in most parts of the world. Virtually all commercial absorption cooling units use the water/lithium bromide solution couple. These have been designed to use steam heating to the generator but recently a number have been modified or specifically developed for use with hot water produced by solar collectors.

Because this is the best known approach this was used as the basis for a first indication of a possible system for the building. It was calculated that with a machine rated at 88 kW and a flat plate collector areas, (based on the Duffie and Beckman double glazed non-selective collector which compares well with a commercially available single glazed selective collector) of 320 m² would be near optimum, and would contribute about 90% of the annual cooling energy requirements. This area of collector would require re-design of the south facing roof, in order to increase its area, compared with that originally envisaged.

The use of advanced collectors was also considered. It was calculated that to achieve the same performance as with flat plate collectors as described above, would require 150 m² of evacuated tubular collectors (based on Corning Glass performance). Alternatively 191 m² of parabolic trough collectors, covering an area of 287 m² (based on performance of Promatic collector) could be used. Energy storage would involve a thermal storage system (water) of about 50 m³ and a chilled water store of about 10 m³.

Photovoltaic Cell/Compression Cycle

As a generator of electricity, the photovoltaic cell has a great attraction in that no thermodynamic cycle or moving parts are involved and no maintenance is required. The use of photovoltaic cells for cooling introduce the considerable advantage that a vapour compression system can be employed with its significant energy efficiency and capital cost benefits.

In order to match the performance of the solar/absorption cycle described above a photovoltaic array of about 10.8 kW (peak) covering an area of 100 m² — 150 m² together with batteries with a storage capacity of about 330 kWh would be necessary. The use of chilled water storage and a reduced battery capacity was also considered. The array could be accommodated on the roof as designed without difficulty.

Aerogenerator/Compression Cycle

Wind energy has been used successfully over many years for small applications, including the grinding of sugar cane in Barbados. In order to give an indication of a possible system, hourly data from Grantley Adams Airport were processed by computer and commercially available aerogenerators were examined.

It was found that, given sufficient storage a horizontal axis aerogenerator of 7.62 m diameter (the Grumman Windstream 25) could supply an energy input similar to the systems described above. As with photovoltaic cells a battery capacity of about 330 kWh or alternatively chilled water storage would be necessary.

COSTS AND ECONOMICS

In general the choice of a particular option for air-conditioning depends essentially on economics, assuming of course that the approach is technically feasible. The three ambient energy based systems described above are technically feasible and their use would lead to lower energy consumption, and hence running costs, compared with the conventional systems. To be economically justifiable any additional capital cost attributable to the non-conventional system should be repaid through fuel savings, within a reasonable time.

Estimates of the capital costs for the three approaches, based on commercial quotations wherever possible, were made. These are presented in Table 1. Costs include the energy collection and storage system, chiller, cooling tower and back up together with controls, plumbing/wiring etc., but exclude the distribution system within the building. Modification to the roof of the building are not included.

TABLE 1

Cost Estimates for the Alternative Systems

System	Estimated capital Cost US $
Solar collector/absorption chiller	
— flat plate collector	138,000
— evacuated tube collector	143,000
— parabolic trough collector	182,000
Photovoltaic cells/compression chiller	161,000
Aerogenerator/compression chiller	108,000

These costs must be compared with the electrically driven compression chiller alone which would cost about US $ 16,000.

The three ambient energy systems described above if employed would displace the consumption of about 16,500 kWh of electricity, compared with the conventional compression system, in a year. At the present price of electricity in Barbados of B$ 0.15/kWh this would be valued at about B$ 2,475 or about US$ 1,300.

Clearly the value of savings could not justify an investment in an ambient energy system. Even the least expensive option would have a straight payback period of 83 years which is in excess of the expected life of the system.

CONCLUSIONS AND RECOMMENDATIONS OF THE STUDY

None of the ambient energy systems are cost effective and so could not be recommended for incorporation into the building on economic grounds. However, it is believed that in the future the use of alternative energy sources will be of major importance and so it may be worthwhile initiating R & D, demonstration and evaluation projects now although they are not presently economic.

Each of the approaches outlined above has a number of advantageous features, as well as displaying some less desirable aspects.

The Solar Collector/Absorption Chiller approach is the one to which the majority of attention has been paid to date, hence sufficient environmental and technical data are to hand to permit a design to be produced, and suitable equipment is available to permit its implementation.

On the other hand the present capital cost is high in relation to potential savings and it is difficult to imagine major reductions being achieved in this type of equipment; scope for local manufacture is limited; the difficulties of integrating the energy collection system into the building are greater than for the other approaches. This approach is already being incorporated into other buildings on the island and therefore it could be beneficial to employ an alternative approach.

The Aerogenerator/Vapour Compression Chiller solution utilising a free-standing wind machine, makes the least impact on the building of all three approaches. It also permits a conventional electrical air-conditioning system to be installed from the outset and — if necessary or desirable — the aerogenerator to be introduced at a later date.

Unfortunately, insufficient wind data exist for the proposed site to enable a thorough design to be completed at this stage and a period of data collection would need to be embarked upon if this solution were seriously to be considered. Although potentially the cheapest approach, it is believed that this approach does not represent an ideal application for aerogenerator based systems; it is suggested that greater benefit would accrue from utilising large aerogenerators located at the most windy sites, to generate electricity for feeding into the national grid. Aerogenerators which are designed to operate synchronous with the grid are now in an advanced stage of development and the study recommended that the Barbados Government seriously consider this means of electricity production for the island.

166

The Photovoltaic Cells/Vapour Compression Chiller solution, although not the least expensive, offers a number of advantages although as with wind, air-conditioning may not be the most appropriate application. It shares the attraction of the solar collector system that sufficient environmental and technical data are to hand to permit design to be carried out, and of the aerogenerator based system that a conventional electric system may be installed at the outset whilst the photovoltaic cell arrays can be introduced later if desired. Additionally, the lifetime of photovoltaic cell arrays is potentially greater than the other two approaches.

Although currently more expensive than the solar collector solution, even more so than the aerogenerator approach, major cost reductions are projected for solar cells in common with electronic devices in general. In addition, the synthesis of systems from large numbers of identical modules makes it possible to introduce the possibility of local assembly of arrays — and even partial manufacture of solar cells — in collaboration with a suitable solar cell company.

In view of the above considerations the final recommendation was that if the Barbados Government decide to proceed with an ambient energy based air-conditioning system for the CARU building, photovoltaic cells should be employed as the means of energy conversion.

Biogas in China

by

Derek LOVEJOY, United Nations, New York

Family uzed biogas digestors, which convert human, animal and crop wastes in a sanitary way to fertilizer values with the simultaneous productioñ of fuel gas for cooking and lighting, has increased rapidly from simple beginnings in 1972. An FAO mission to China in May 1977 reported some details of this programme and at that time there were to be about 1 million units in operation, mostly in a few countries in the Southern part of China.

Such has been the growth of the biogas movement that by the end of 1978 more 7 million units were said to be in operation including many in the north where climatic conditions restrict their operation to 8 or 9 months of the year when the temperature is above freezing. With approximately 150 million rural families in China, little more than 5% now have biogas but such is the rate of introduction that this percentage is expected to increase rapidly in the next few years.

To understand the undoubted popularity of biogas digestors it should be recalled that agriculture in China has been, over many countries, land-intensive in a way, unknown until recently in most other parts of the world. In particular this has required the meticulous recycling of organic wastes, human, animal and vegetable, in order to maintain the fertility of the soil. The regular collection of "might-soil" and its distribution as fertilizer has for centuries been practised in China.

When asked what were the principal benefits of biogas, one farmer replied immediately that it freed his children from the laborious (and unsanitary) taks of collecting might-soil so that they could spend more time with their school work.

Because of the primary emphasis on sanitation and fertilizer, Chinese biogas digestors are designed for simple and cheep construction and operation without regard to efficient gas production. One, more or less standard, design has an underground chamber with a domed roof which collects the biogas together with inlet and outlet tubes. Hydrostatic pressure results from the difference in liquid levels in the inlet and outlet tubes over that in the main chamber. With a volume of about 13 m^3 some 5 m^3 of biogas (60-70% methane with the balance mostly carbon dioxide) is produced daily in warm weather. With a calorific value of about 20 MJ/m^3 the gas is sufficient for the cooking and lighting needs of a Chinese family of five.

The input to the digestor is typically provided by an outside latrine separated by a particular wall from a sty containing one pig (family owned). In addition to human and animal wastes, approximately equal quantitites of straw are added after composting the latter for at least 10 days and diluting with water to reduce overall solids to an optimal 7-9%.

In addition to the biogas, the digestor yields each year some 15 m^3 of effluent (one analysis showed 500 ppm available N, 15 ppm available P, 2 000 ppm K) and 10 m^3 of solid sludge (somewhat richer in fertilizer values) which latter must be scraped out every six months. Sale of fertilizer to the commune yields the family some 50 Yuan (USD 30) annual income.

Construction is based on traditional and local materials. In one widespread version the walls and floor are made of traditional clay/lime (95%/5%), thinly plastered with cement to ensure gas tightness. The dome and inlet and outlet tubes are of cement, and the dome has a manhole for access (sludge removal) and a polythene tube to transmit the gas to the kitchen stove and mantle (lighting). A simple water U-tube measures gas pressure.

A training school in Sechuan province (south central China) runs four week training courses which include three weeks of classroom and one week to build a digestor. Graduates of this training course then hold training courses at province and country levels for trainees selected by each brigade (village) whith typically 100 or 10 families.

Each brigade then has a paid biogas technician who designs and supervises construction of digestors and assists in their subsequent maintenance.

Construction is typically carried out in winter and slack times with five families cooperating in the simultaneous construction of five digestors under the supervision of the technician.

One biogas digestor requires about 1 000 kg of lime and 100 kg of cement and takes some 20 man-days of self-help labour. Each family puts up half of the 80 yuan (USD 50) cash cost of the digestor and receives an advance of the other half, which is repaid within a year from sale of fertilizer.

In a particular commune near Beijing (Peking) per capita cash income was said to be 156 yuan annually (600-800 yuan per family) so that the outlay was not excessive.

While present emphasis appears to be on privately built and owned family size digestors, a larger village size digestor supplying a suitably adapted diesel engine and generating 15 kW of electrical power for the village was reported under test in Sichuan.

Another experimental and development work was reported to be proceeding at a number of centres including microbiology studies on higher yielding fermentation bacteria.

Future Energy Strategy and Solar Energy Utilisation in Papua New Guinea

by

Dr. ASHFAQ AHMAD

University of Technology, Papua New Guinea

INTRODUCTION

Papua New Guinea is an oil importing third world country, entirely vulnerable to temporary curtailment of exports from the oil producing countries (OPEC), along 1973-74 lines, and to the rapid growth in price, in real terms of oil products relative to other goods and services which will certainly take place during the 1980s and thereafter.

On conservative estimates firewood makes up 39% of total energy demand. Petroleum products sales amount to 4 million barrels of oil per year, and represents 59% of energy use. All fuel oil, and about one third of diesel, is used in industry, and practically all petrol is used for transportation. Firewood, other than for steam generation in the timber industry, is entirely a domestic fuel, more than 90% of which is used in village communities away from urban areas. Electricity is largely an urban energy form as rural electrification is virtually non-existent.

It is obvious that almost the entire industrial basis of the national economy, including the marketing of rural produce, and the distribution of goods and services, is critically dependent on imported petroleum products, and at a time when the future economic and strategic stability of the supply of these commodities is in considerable doubt.

A parallel and interrelated crisis in the supply of energy in Papua New Guinea is the demand for firewood. Until recently little has been known of the dynamics of firewood supply and end-use in PNG, but recent research in Simbu and Morobe Provinces is enlightening.

In the Sinasina region of the Simbu Province average per capita consumption of firewood is now 1.2 kg/capita/day. This level of consumption is the lower end of global comparative figures close to India's National average consumption, and below that of Nomadic people in the Sahel Desert. It is the same level of consumption as settlers in Lae, who despite not needing firewood to keep warm, still find it difficult to keep up firewood supply. There is a regular trade in firewood at local markets at 4t/kg, and 6-8t/kg in Kundiawa. Not only is firewood gathered from a 4-5 mile radius, but those that can afford it hire PMVs to bring in special forest timber (Nothofagus and Castanopsis) for the walls of their houses. Trees are often sold for K30-K40 each.

The relationship between this problem and the oil crisis is obvious and has parallels in India and Malaysia. Once firewood supply is short, kerosene becomes the only alternative. Money derived from cash cropping is partly channelled to stoves, lamps and kerosene fuel, and this despite a price of 20 t a coke bottle in the Simbu

Province. Again only the village elites participate, but to the extent that they can, and do, they become hooked on imported kerosene of uncertain supply, and rapidly increasing price.

In summary the supply of firewood is not a separate problem but yet another part of a major energy crisis now emerging in PNG. And neither is the problem one of energy supply alone, for the deforestation of rural areas, in turn, increases soil erosion, and destabilizes subsistence agricultural productivity. Similarly, the time spent gathering wood eventually reduces that available for other subsistence activities. The afforestation of Goroka which began in the 1950's is testimony that replanting programmes can work in PNG, and justifies greatly increased support for the Office of Forests which initiated these vital programmes.

FUTURE ENERGY STRATEGY

Using as a basis for estimation of future demand detailed data on energy use in the major industrial centre of Lae, the following scenarios are presented. All represent energy growth without specific government intervention, although it is likely that the "no-growth" and slow-growth scenarios will not be fulfilled in the short term without policy intervention.

The three scenarios are:

(1) **No-growth scenario.** No increase in per capita consumption of energy which currently stands at 23 MJ/capita (0.29 kW), and 16 MJ/capita (0.19 kW) for firewood. Only population growth will drive total energy demand.

(2) **Slow-growth scenario.** 3% growth in per capita energy demand. The rate at which GDP is anticipated to rise, and the rate at which commercial energy demand has grown in the last three years in PNG.

(3) **Maximum-growth scenario.** 8% growth in per capita energy demand. This is the rate of growth in commercial, or imported energy forms in PNG during the last decade, and is the rate of growth that might be expected if the economy strengthens to once again parallel the average growth in GDP in oil-importing countries in the region in the period 1960-73.

In all scenarios allowance has been made for urbanization of the population, as well as total population growth. For most petroleum fuels the current urban to rural ratio in per capita consumption is about 20:1.

Even with no increase in per capita consumption there is anticipated to be a 2-3 fold increase in imported petroleum products by 2000. With the more realistic slow-growth scenario, of 3% per capita energy growth, we expect a 5-fold increase in imported energy forms by 2000. If there is strong economic development during the next 20 years, however, we can expect as much as a 15-fold increase in total imported energy demand. This level of increase to the year 2000, as extreme as it may seem, will still leave Papua New Guinea with a per capita consumption below the world average per capita consumption of today.

Unless we implement widespread and quick-acting conservation measures now, forward planning on the basis of the 'slow-growth' scenario will lead us to under-estimate near term growth in total energy demand.

It is obvious that the proportion of total foreign exchange used on imported petroleum products is increasing rapidly, and currently represents an expenditure of close to K80 million (1978 Kina). It is not inconceivable that by the early 1980s energy imports will demand 20-25% of total foreign exchange. Already it is clear that there are major financial incentives for PNG to move rapidly towards the 8 point plan objective of self-reliance in the energy sector, and thus to implement demonstration and development programmes as soon as energy options are agreed.

Finally, it is obvious now that the supply of energy has become a significant economic burden to the nation, and that the security of present energy sources necessary for stable economic development is outside the control of the national government. What economic value is attached to a secure energy supply is on one hand an academic question but on the other it raises real questions of the level at which a government will support an indigenous renewable energy development programme.

Thus for the realisation of the 8 point plan objective of self-reliance in the energy sector, the following general points should be considered for formulating the policy for meeting energy demand from renewable locally-available energy sources.

(1) The development of locally available renewable energy sources will be a significant move towards self-reliance and political independence.

(2) Monies saved from direct purchase of fuels on the world market will not only save foreign exchange but will represent monies spent within Papua New Guinea in support of Papua New Guinean owned industry.

(3) Most alternative energy sources will prove economic first in rural areas where delivered costs of imported fuels are high. Energy industries developed in rural areas will help fulfil the policy of balanced rural development, and will give effect to self-reliance on a regional basis.

(4) The alternative energy industries can all be geared for significant smallholder participation, fulfilling the goals of improving the participation of nationals in the economy, and increasing equity in the participation in national economic growth.

The strategy for substituting petroleum products in industry with renewable indigenous energy sources recognises that in the short term we must produce energy forms compatible with combustion in existing energy-using equipment. Later on, as new industry is established in PNG or as existing industry expands, energy conversion equipment specifically built for the efficient use of renewable energy forms can be installed.

The Bulk Generation of Electrical Power from Solar Energy with Thermochemical Heat Transfer for Developing Countries

by

M.S.A. SASTROAMIDJOJO, Indonesia

In view of the need for pollution free energy sources of which solar energy is the most globally available, it must be realised that domestic use of energy forms only a negligible part of the total national energy consumption. In Indonesia this would amount to about 0.4% of the total energy use.

This makes it imperative to direct the national effort to solar electricity power plants with capacities of multi-megawatt scale for industrial use.

The means to do this are now at hand.

I do not want to start an overview of all the bulk electric solar power schemes in the world today, like that of the power tower concept in the USA or the flat plate small scale power plant (in the tens of kilowatts) being tried out in West Germany and France.

Instead I would like to talk about the engineering and economic reasoning for arriving at a solar energy power station competitive with the OPEC oil cost on the open market.

It is only when the system as a whole is competitive with fossil fuel systems of today that there will be wide spread acceptance of solar energy.

— What elements in a solar system form the most important part?
— What kind of technology is needed?
— Is there scope for international co-operation and development of large scale solar power plants?

I suggest that the answers to the above questions are as follows:

What elements in a solar system form the most important parts?

The collectors. A sun-tracking paraboloidal mirror is an obvious choice. For a paraboloidal sun-tracking concentrator the concentration ratio is the highest, e.g. between 5 000 and 10 000, depending on the degree of perfection in manufacture, while the equilibrium temperature can be as high as 3 000 °C-3 500 °C.

High temperature collector development combined with appropriate heat transfer and storage is the key for using high temperature thermal fluids in the conversion of solar energy to mechanical work (pumps, engines for driving generators etc.). Again, thermochemical heat transfer and storage is most efficiently done with high temperature collection. High temperature chemical exchange processes for the direct production of hydrogen may also be realised directly.

It is with the above in mind that the SERC (Solar Energy Research Centre, Gadjah Mada University) has concentrated on ways and means to lower the price of paraboloidal mirrors. The latest results at SERC indicate that a thin (3 mm) reinforced concrete shell, with appropriate reflective coating is a definite possibility. In the same context a study was made of the thermal distribution in the focus of concentrators such as paraboloidal mirrors. It is hoped that in this way more effective focal absorbers can be developed.

Much ongoing research is directed to the realisation of cost-effective paraboloidal collector arrays, together with methods for construction, reflective surface production, control and protection.

What kind of technology is needed?

A measure of sophistication is the ratio of the cost of a product to its weight and on this basis bitumen roads are of the order of $ 50/tonne while computers are of the order of $ 50 000/tonne.

The maximum allowable solar system cost can be referred to collector area by relating the volume V of material in a solar collector system to the collector area A. If $V/A = t$, and ρ is the material density, the allowable sophistication S (in dollars/tonne) can be given by:

$$S = \frac{C}{\rho t}$$

One can make collectors resembling brick walls or paper thin solid state collectors. In between these extremes, keeping transport costs in mind, the most favoured option would be that of thin (in terms of t) self-supporting structures, for example paraboloidal dishes.

For a paraboloidal dish of $t = 3$ mm and $S = \$ 3\,000/m^3$, the costs would be $ 100/m^2 and possibly less when manufactured on a mass production scale. T_c moreover would be of the order of 500 °C which would mean higher thermodynamic efficiency.

I will conclude with the observation that a disporoportionate amount of time, money and talent is apparently being expended on 'academic' research. We have seen this in the development of selective surfaces. Creative thinking should be directed to the psychological economic and engineering problems of why and how solar energy by and large is not yet accepted.

Finally the solar energy proponent should shift his stance from "fossil fuel depletion" incentives to "environmental" considerations for the simple reason that the environment will be irrevocably damaged long before non-renewable fuels are exhausted.

Is there scope for international co-operation and development of solar power plants?

Definitely, but not the way in the ocean big fish often co-operate with small fish.

Many developing nations are situated in tropical areas where there is both abundant sunshine and abundant rain.

Large scale solar power plants for "reconnaisance in force" in those areas will give a wealth of data needed for future design in other parts of the world.

174

New Forms of Energy in French Polynesia

by

Mr. JOURDE, France

A. GENERAL

The programme for developing New Forms of Energy in French Polynesia is in two phases.

First phase: Demonstration — 13 test operations between 1978 and 1980.

Second phase: Creation of a local industry.

— This programme was launched at the request of the Territory of French Polynesia (which has a statute of internal autonomy). An overall enquiry into energy problems in Polynesia had previously been carried out by the CEA.

— It is financed by the Territory of French Polynesia on the one hand and by SECOM and the CEA on the other.

— The overall cost of the test programme amounts to 3 million dollars, or 1 million dollars per year between 1978 and 1980.

— The ultimate goal is to redress the balance between Tahiti, the main island (consumption: 1.5 TCE per capita per year), and the hundred or so islands and atolls scattered over 4.5 million km² (consumption: 0.1 TCE per capita per year).

B. FIRST PHASE — TEST PROGRAMME (1978-1980)

This programme is aimed at on-site testing of the various answers provided by wind and solar energy to the needs of the islands and atolls: Cooling - Water - Electricity.

The programme includes the first three years of running as well as putting various measures into operation at these installations.

1. Evaluation of wind and solar resource

5 stations located since 1978 in each of the archipelagoes to supplement the meteorogical network, in particular for solar measurements (global and diffuse radiation).

2. Cooling production

On an industrial scale:

— 1 fish storage unit: 1979
10 m³ tank of sea water kept at − 2°C and taking
1 tonne of fish per day

— 1 fish freezing unit: 1980
300 kg/d - 20 m³ storage chamber at - 18°C

— 2 buildings with solar air conditionning   1979 and 1980

On an individual level:

— 5 types of individual refrigerators:
Compression + photovoltaic cells
Peltier effect + photovoltaic cells
Compression + wind   1978 to 1980
Absorption + concentrated solar
Zeolite + concentrated solar

3. Water

— 1 unit pumping fresh-water by photovoltaic
cells   1978

— 1 unit pumping fresh-water by multiblade wind-
power generator   1978

— 8 stills of hot-box type (manufactured locally or
imported)   1978-1979

— (1 reverse osmosis unit + wind generator)*   1981

4. Electricity production

— 1 18 kW windpower generator supplying a grid   1979
— 5 separate habitations — lighting, radio, TV, etc.
powered by photovoltaic cells or windpower   1978 to 1980
— (gas generator 20 kVA)*

5. Tests of equipment and components under local conditions

Tests of water heaters, photovoltaic cells, concentration systems, in the very severe climatic conditions of the atolls.

---

* Complementary operations not currently established.

C. SECOND PHASE: LOCAL INDUSTRY

Date of going into service

— Training of Polynesian management and technicians in universities and research centres — 1978 to 1980

— Creation of a local industry
  - Manufacture of water heaters and various other components — from 1978
  - Engineering office — 1981

Non-Conventional Energy for Rural Needs in Pakistan

by

SHAHID IKRAM & M. M. QURASHI, Pakistan

INTRODUCTION

Economic development today is highly dependent on energy which is required to produce the necessities and amenities of life like food, shelter, communication, transportation and so forth. In fact economic development consists in large part of harnessing increasing amount of energy for productive process. This could only be possible either by tapping alternative sources of energy or by making more efficient use of available energy sources through the use of appropriate tools and conservation techniques.

The developing countries, like the rest of the world, seem involved in a continuing struggle with increasingly high relative prices for energy from conventional sources. The 'energy crisis' will continue to be a fossil fuel crisis. This conclusion should not be surprising, because evens in the industrialised natio, with their infinitely greater capacity of research and development, capital accumulation, skilled manpower and ability to take risk, at least a decade or two must pass before any of what are now recorded as promising unconventional energy sources will play an important role in the energy supply picture. Even before a particular energy technology holds definite promise, the prospect of the short term benefits to developing countries is doubtful; because these sources do not match well with the demands of energy users in developing countries/nations.

Increasing energy supply and efficiency of energy use would be important aspects of any such comprehensive development strategy.

To cope with this situation the promising energy techniques involve the utilisation of wind power, hydro-electricity and solar energy for various purposes. The prospects of developing new energy techniques are bright because of the price hike of fossil fuels and expensive nuclear power sources.

The current potential value of unconventional energy techniques for use in developing countries like Pakistan has been examined, and this paper gives a brief account of the present position with emphasis on small scale application suitable in rural areas. The most promising current uses involve mini-hydel, electric power from solar energy, solar heating, solar cooling and water pumping with wind power. Research may be turned to reduce the cost of electricity.

RURAL ENERGY PROBLEMS AND POSSIBLE SOLUTIONS

Most of the alternative energy technologies are at the moment inaccessible to most rural communities in developing countries. This is not because of the lack of skills in the village, but because of the lack of the capital to acquire necessary material and tools. Education training, however, is required; indeed the village or farm dweller usually understands his problem better than other do, and if given necessary financial and material resources, he can adopt technology more effectively than most outsiders are willing to believe. A specific rural area should be selected for study, as an economic unit, by a team of technologists and rural development economists. The purpose of this on-site study would be to identify methods to advise an appropriate scheme consistent with material development objectives, to enable the chosen areas to asopt appropriate alternative energy systems. This could be accomplished by providing training and information to the people from academic community and others who are responsible for project implementation in the field, including village leaders. This would enable the community to make more efficient use of its human and material resources.

The major contribution of energy to economic growth in Pakistan and others less developed countries lies in the areas that require conventional kinds of energy and this probably will continue to be the case at least the next few decades with no easy relief on the side of supply. Although the promising 'non-conventional' energy technologies often relate to demands that are matters of comfort (light, radio, TV), there are applications of these technologies that can indeed contribute to overall economic growth, e.g. solar crop driers, gas from biomass or solar cells for communications.

THE ATDO'S CONTRIBUTION TO RURAL ENERGY DEVELOPMENT

A short résumé of ATDO's and other developmental work on "non conventional" sources of energy, is given below:

1. Hydro-electric Installations:

In the northern areas of Pakistan, there are a number of small streams and natural falls which have sufficient potential hydro-power for development of micro and mini hydro-electric plants. The ATDO have in hand a scheme to help the people to have the supply of electricity in the far flung areas specially of the northern part of Pakistan at cheap cost with reasonable standards of voltage and frequency. The inaccessability of these areas render the proposition of connecting them with the main network of electricity supply unattractive at this stage. Development work proceeds on two lines; one on simplified pelten wheels, on sites having medium to high heads and small discharges where wooden wheels may not withstand the force of water; and the other on wooden wheel or wheels made with PVC for use of low heads and large discharges, where usual grinding mills are being operated. After designing, the turbines are fabricated and tested in laboratory and then installed on selected sites. The ATDO hydro-electric installation is broadly divided into 2 categories:

(a) Micro Hydel Plants: (1 kW approximately) capable of manufacture/assembly in villages.

(b) Mini Hydel Plants (20 kW to 200 kW) requiring town machineshop facilities for manufacture.

2. Wind Energy

The exploitation of wind power holds great promise, especially to initiate development activities in areas where there is scarcity of water, by providing much needed water for the settlement of scattered communities which roam about in search of water. Suitably placed windmills can transform the desolate landscape into a prosperous and happy area resplendent with cattle, sheep and geat, raising-farm and cottage industries. A windmill no doubt has a very low efficiency of conversion, yet the input power being without cost and inexhaustible and manufacturing being relatively inexpensive, it could have an edge over others. The versatility of a windmill and its ease of erection and maintenance renders it as an ideal tool to lift water in areas where there is scarcity of water, for agriculture and cattle raising purposes and for lighting and small power needs.

The main constraints preventing wider use of windmill in Pakistan are:

(i) The relatively high cost of locally fabricated windmill.

(ii) The existing manufacturers are unwilling to incur the expenses of R & D and retooling produce more cost effective designs.

(iii) Lack of expertise and experience of manufacturing the windmills to suit the local wind conditions and pumping requirements.

It is, therefore, concluded that suitable windmills designs are needed which would conform with comparable outputs and reliability to existing commercially available equipments, but which would be less material intensive and involve production techniques that are already widely available in small and medium-sized engineering workshops of a developing country.

Although windmills can also be used for generating electricity, no experiments have been done so far in Pakistan to obtain regular power supply from the windmills.

3. Bio-gas Technology

The anaerobic fermentation of organic materials in a closed vessel with a narrow outlet for gas, produces a sufficient amount of gas which can be used as fuel. The fermented slurry is a better quality fertilizer because in the process of fermentation, the bacteria adds up to three times the quantity of nitrogen already available in the dung. Nitrogen thus improves the quality of manure. This technology suggests a good alternative source of energy which can bring a new dynamism to the entire village economy. The gas besides cooking can be effectively used for lighting and power generation for pumping irrigation water, grinding fluor, etc. While the remaining manure can be used as a very effective fertilizer. The Energy Resources Cell Islamabad, Pakistan has put up as many as 54 Bio-gas plants based on Indian technology, mainly on dairy farms.

In China, a simple Bio-gas Plant has been developed. Being a brick structure with a dome type roof, it has an added advantage over Indian type plant where a steel gas holder is a sine-qua-non, while it has almost no requirement of steel and therefore, can be constructed at a comparatively cheaper cost. The Chinese Technology was,

180

therefore, obtained and it was decided to construct a few plants at Government expense to judge the effectiveness and suitability under Pakistan conditions. Some useful changes heve been introduced in the original design. For example, the mixing of dung and water is done before the slurry is added to the digester tank, in separate tank over the inlet. Previously, both the ingredients were separately put in tank and then stirred which was **not** so effective. It has been our experience that instead of drying the slurry to solid form, if liquid slurry is applied directly in small dozes to the plants, over a period of time, it is more beneficial to the growth of vegetation.

To make the effective dissemination of this technology in the rural areas the Pakistan Banking Council was approched to provide loan facilities to the small entrepreneurs for installation of these plants. Realising the importance of the scheme, the Banking Council had directed the Commercial Banks to provide not only loan facilities to the intending users of the technology but have instructed them to act as technical know-how despensing centres.

In additon to the 10 m³ units, a 50 m³ is being installed at Tando Jahanian, Hyderabad. The owner intends to use the gas as fuel for an engine which will be use as prime mover for fodder chopping machine. Another similar plant is being installed near Karachi for the same purpose.

4. Direct Solar Energy

Out of the various alternative sources of energy, direct solar energy has possibly got the greatest potential for future uses. Solar radiation by earth is to the order of 1.2×10^{14} kW, which is approximately twenty thousand times, the current world energy requirement can be tapped in near future at two levels:

1. At Sophisticated level
2. At low level of Sophistication (small-scale application).

In the first category fall the following technologies, which are costly and large-scale oriented:

1. Electricity generation.
2. Cooling of Buildings.
3. Large Scale Solar Desalination.

In the second category fall the following technologies, which are simple small-scale and low-cost:

1. Solar dehydrators for fruits and vegetables and crop drying, etc.
2. Solar desalination.
3. Water heaters and space heaters.
4. Solar pumps.
5. Solar Power Generation.
6. Solar lighting for Rural Areas.
7. Solar Cooking.

The ATDO, P.C.S.R. and the Energy Resources Cell are currently devoting themselves to adapting and field testing small-scale solar technologies, since these technologies are readily available and require relatively little R & D work. A few projects in which work is being done in Pakistan are as follows:

1. Dehydration of Fruits & Vegetables

A lot of fruits and vegetables go waste in the country due to lack of storage and preservation. Dehydration is a way to preserve food. The ATDO initially assigned work to some local Organisation to develop a low-cost cottage-level design for a dehydration plants, but ultimately the ATDO had to do their own work, based on technical knowledge collected from all over the world. Partial success has been achieved by having a small UNICEF designed dehydration plant which is simple enough to be operated by villagers and mostly uses solar energy. A demonstration model plant was fabricated and operated by ATDO for a few months and results are satisfactory.

2. Solar Desalination

The use of solar energy for desalinating seawater and brackish well-water has been demonstrated in several moderate size pilot plants in several countries. This basin type century old process has been modified and adapted to modern materials. It consists of a shallow pool of brine from which a slow evaporation of water takes place. In its modern form it is covered by slopping sheets of glass, water condensing on the underside of the cooler glass covers, runs to throughs at the lower edges and to storage. Excess brine that has not evaporated tun to waste as salt water is supplied to the basin. On such large scale unit has been installed by the PAEC at Gavadar, Baluchistan.

3. Solar Water Heaters

Solar water heaters constitute a fairly recent development based on a common natural phenomenon. Cold water in a container, exposed to the sun undergoes a rise in temperature. In its modern form, the solar water heater is basically a flat-plate collector and an insulated storage tank. The collector is commonly a blackened metal plate with attached metal tubing and is usually provided with a glass cover and a layer of insulation beneath the plate. Developmental work on these is being undertaken at P.C.S.I.R. Labs, Lahore.

4. Solar Water pump

Many countries are using solar energy for water pumping. PAEC provided the following details.

A solar water pump was designed which utilizes a 21 ft. diameter paraboloid fitted with flat mirrors, boiler and a steam engine. The complete system was made with locally available materials and it provides 2 H.P. to the water pump. In order to study smaller portable units for 1/4 H.P., five ft. dia, mosaic mirrors were constructed and connected to steam engine and boiler. University of Engineering & Technology is also working on this project.

5. Solar Power Generation

Solar energy for power generation, either in the form of electricity or mechnical work, has been the subject of extensive research in many countries. Much of the work has been directed toward the use of reflecting surfaces to concentrate the solar energy into a small receiver/boiler, which permits the development of much higher temperatures than is possible with flat-plate collector. High pressure steam for electric power generation or with precise equipment, extremely high temperature heat for chemical and metalurgical processing can thus be produced. Concentrators in the form of parabolic cylinders (troughs), and other shapes have been investigated. An important alternative is the use of solar cells which are expensive, but technology for making which is constantly being improved and needs to be watched by developing countries with an eye to the future. The E.R. Cell in Pakistan is in the process of implementing four village demonstration projects of 1-2 kW, under a U.N. programme, for photovoltaic power generation with Biogas back-up.

Irrigation Utilizing Solar Energy and Bio-mass and Solar Powered Lift Pump

by

Prof. MUTHUVEERAPPAN, India

ABSTRACT

Rural areas and agricultural fields lack conventional forms of energy in many places. It is highly desirable to use solar energy for lifting water for irrigation or to meet the daily requirements of a small community. This paper deals with initial stages of development of two types of pumps developed at the Annamalai University, India.

Irrigation lift pump utilizing solar energy and bio-mass will consist of a flat plate solar collector, a concentrating type solar collector requiring little orientation, bio-mass fired auxiliary energy booster and a simple pump of robust construction with no moving parts except for two non return valves. This system will operate on steam produced by solar energy and bio-mass heating with continuous supply of steam for its working. A small experimental pump has been fabricated and tested using steam which in this experiment is being supplied from a small boiler for the feasibility study of its working.

Solar powered lift pump differs from the pump discussed above from the way in which steam is produced and the pump is operated with continuous supply of heat. Concentrating type of solar collector with high concentration ratio is required to collect the solar energy. This energy is transferred to the pump by using heat pipes. The pump consists of an evaporator unit and a condenser unit connected by suitable pipes. Steam is produced inside the pump itself and utilized for the pumping operation. A model of the pump has been fabricated and tested for its working with kerosine stove heating in order to study the pump operating characteristics.

INTRODUCTION

World is now very conscious of the scarcity of conventional types of energy and the price hike of fuel oil and coal makes one to think of the importance of alternative sources of energy. In countries like India where there is abundant sunshine. Solar energy can be easily and effectively utilized. India is primarily an agricultural country and most of the crops depend on irrigation that too on lift irrigation. It is estimated that more than 2 million tonnes of bio-mass like husk, bagasse, groundnut shell etc., are available in India. Hense these simple pumps utilizing solar energy and bio-mass will be highly suitable for remote villages.

WORKING PRINCIPLE

The pump body has been fabricated by selection of suitable pump barrel, suction line and delivery line. There are two non return valves one at the suction line and another at the delivery side. Wheel valves for control of steam supply to the pump body and for the venting operation are provided. To begin with the pump barrel is filled with water so as to remove air from the system. Steam produced in a small boiler is supplied to the pump continuously and it has been found that water is being lifted and indicating the feasibility of its working.

Steam when flows at a particular velocity it strikes the water surface inside the pump body and creates turbulance so as to initiate condensation in the pump barrel. Initial condensation creates vacuum which sucks the water into the pump body. When once fresh water enters into the pump body vacuum is agumented very rapidly. This creates a rapid flow of water in the suction line. This high velocity of water when suddenly brought to rest by the on coming steam, high pressure is developed as in the case of hydraulic ram. This forces open the delivery valve and the water is delivered at a faster rate while the water flow in the suction line is being continued due to inertia. Water flows from the sump through the suction and delivery valves till this pressure dies out. This is similar to the automatic action of the hydraulic ram. When the inertia forces die out, the oncoming steam now pushes a small quantity of water from the pump body through the delivery valve by virtue of steam pressure. This water is heated and thereby the heat of condensation is thrown out. The second cycle starts by the initiation of condensation of steam.

It is also noted that the ram effect is having a large quantity of water delivery and also a small quantity of slightly warm water possibly because of the inertia effect dies out in succession. All along the steam supply is maintained at the same pressure and at the same quantity. There is no steam cut off and there are no ports in the system. The pump works similar to hydraulic ram and hence it may be called a thermal ram pump.

The pump barrel should not be lagged as otherwise condensation could not be induced. The rate of steam flow and the rate of steam condensation through the pump body surface play some role in the pump operation. The exact location of condensation of steam in the pump body varies with the parameters such as the steam pressure and the rate of steam supply velocity of steam flow in the pump body is also an important parameter which is to be optimised.

EXPERIMENTAL STUDIES

In our experiment we have tried with varying supply pressures and steam rates to study the pump operation. The steam pressure need not be very high. The pump is successful even at low pressures in the range of 304 kpa to 1,013 kpa. The water pumped through a head of 0.8 m is found to be at the rate of 0.214 Kg/s. consuming a total steam supply of 0.0041 Kg/s. of steam. The pressure in the small boiler fabricated by us could not be maintained at the desired pressure while the required steam is supplied. Hence pressure was falling from 1013 kpa to 304 kpa while water was pumped at an average rate of 0.214 Kg/s. When the pressure was below 304 kpa the rate of pumping was less. At the start with 1013 kpa steam supply, the water pumping rate was high and if sufficient steam is to be supplied continuously at this pressure the total water pumped can be very high consuming less quantity of steam.

In the present investigation it can be said that this new type of pump is feasible and 50 Kg of water can be pumped for every Kg of steam produced. The head of pumping in the model is low but a few trials with larger head have also been tried and it shows that the pump can work for larger head too.

There is no other fluid other than the pumped water as the cooling medium for this type of steam engine to discard the unused energy. Further steam is the working medium in this pump and the pumped water is the coolant. There is no prime mover like steam engine or steam turbine. Even if there is leak in the system engine or water can be replenished. The efficiency of the combined engine and pump comes to 0.015% but it is strongly felt that it can be improved further so as to have a possible efficiency of 0.15%.

The solar and bio-mass heating systems can be designed and tested after exactly assertaining the parameters which affects the working of this simple pump. When once steam supply rate and the pressure of steam are determined for the given pump the solar and bio-mass heating can be designed to test the entire system.

SOLAR POWERED LIFT PUMP

Since the solar powered lift pump is of low horse power unit, the steam producing unit and the pump unit need not be separate entities as the steam supply required is very small and no continuous steam srupply is required. Combining the steam producing unit and pump solves many problems. This pump differs from the pump discussed above from the way in which the steam is produced and the pump operates. Further this does not require steam supply from an external source.

LIFT PUMP SYSTEM

Paraboloidal concentrating type of solar collector can be used to collect the solar energy and this heat energy is to be transferred to the evaporator unit of the pump through a heat pipe. The heat pipe is necessary because of the following reasons. 1. The heat can be focussed only at the bottom of hte absorber (target) but here the heat has to be supplied at the top of the evaporator). 2. To transfer the heat from the solar collector to the evaporator section of the pump. 3. The system permits the location of the pump at convenient place. The solar concentrators though having tracking difficulties have higher collection efficiencies thus resulting in smaller collectors. To achieve higher temperature and higher collection efficiency, paraboloidal collector is preferred to produce steam. The heat pipe has been selected to transfer the heat from the paraboloidal collector to the evaporator unit of the pump as its efficiency of heat transmission is very high.

CONSTRUCTION AND PRINCIPLES OF WORKING OF THE LIFT PUMP

The pump consists of H type pump body with an additional link at the top. The two vertical legs are of larger diameter pipes and the connecting links are of smaller diameters. One of the vertical pipe acts as an evaporator unit and the other as condenser unit. The two non return valves which are connected to the top and bottom of the condensor legs with suitable heads are the only moving parts in the pump. First the pump is primed to remove the air from the pump body. A laboratory model was fabricated and tested. Then the steam was produced by heating the water at the top of the evaporator by using a kerosine pressure stove. The steam produced pushes the water through top horizontal pipe and opens the delivery valve. As soon as the steam reaches the other vertical leg the condensation takes place. The condensation produces vacuum by which water is sucked inside the pump body through the inlet valve. The sucked water acts as the coolant and fills up the evaporator leg completely through the bottom connection. Then the cycle is repeated by heating of the sucked water in the evaporate unit. The operation of the pump is intermittant though the heating is continuous.

CONCLUSION

Flat plate solar collector and a simple concentrating collector requiring little orientation will be suitable to supply the necessary steam. Bio-mass heating can be used as an auxiliary heating arrangement. The system developed should be easily serviceable by the rural public hence the solar energy concentrators requiring automatic orientation is not recommended. Further an auxiliary energy supply is desirable for any solar appliance to guarantee a reliable and continuous operation. For these reasons the use of flow grade bio-mass, agricultural wastes which are available at throw away prices in comtemplated. This boost up the energy level and itensifies the steam the single fluid system with continuous steam supply without a prime mover such as an engine or turbine will be well suited. This unit which is the combination of the prime mover and pump is unique in nature.

The overall efficiency of the above systems is no doubt lower but this will be outweighed by the simplicity and the inexpensiveness of the system. By proper design the overall efficiency of large scale unit is expected to improve.

Solar Heat for Agriculture

by

Prof. MUTHUVEERAPPAN, India

INTRODUCTION

This paper is on Solar Energy for Agriculture. India is a vast and basically agricultural country. About 80 percent of her population are living in villages. The rural public are not generally benefitted by the modern energy source, namely electricity, cooking gas, fuel oils etc. Solar energy is available in plenty at no cost particularly in tropical countries like India. We can introduce easy and scientific methods to utilise solar energy and wind energy in villages to meet their day-to-day energy requirements.

In villages, farmers utilise solar energy in a primitive way for crop drying, fodder drying etc. Solar drying can be greatly improved by introducing solar falt plate collectors, wind, animal and solar operated draught systems. Further to improve village life we can introduce solar stills, cabinet dryers, water heaters, cookers etc. for household applications.

The best way to engage the farmers during the agricultural off season is to start cottage industries. There are conventional energy saving devices such as solar palm juice concentrators, solar distilled water plants for lead acid batteries, solar wax melter, solar tea and coffee maker, solar drying units for textile factories in villages. Small commercial solar dryers for producing pickles, papadam, dried vegetables, sago and rice products are to be developed in order to save fuel energy, time and to minimise losses and contamination.

Commercial solar crop dryers for tea, coffee, tobacco, chillies, groundnut, pepper, fruits, fish etc. can be introduced to improve the quality of the produce and to save time in marketing. Timber seasoning requires solar energy utilisation for better quality and early marketing.

Equally wind power can be utilised for pumping, providing draft for drying, generating electricity for lighting and battery charging for the village development.

SOLAR DRYING

Solar drying of agricultural products in direct sun on ground or courtyard or even on roads is a common practice in almost all villages in India. In this process, usually high labour cost, inferior quality because of uneven drying, over drying, contamination by insects, losses due to birds, animals etc. are frequently met with Conventional floor drying involves risk of rain damage due to sudden down pour. Improved solar drying can eliminate the losses and solve our food problems.

A solar paddy drier of about 50 sq.m. area was successfully developed by the Annamalai University, India. This project was funded by Department of Science and Technology, Government of India. This dryer is a flat plate collector unit. The roof of the building is the flat plate collector itself and this reduces the cost of the system. The building is available for storage of grains or for process machinery. Cost analysis has shown that this solar dryer is cheaper in comparison with conventional dryer using fuel oil. It is possible to get hot air at about 55°C with a solar collection efficiency of about 50 percent. This 55°C air is sufficient for many grain drying operations. Hot air removes the moisture from the grain quickly and thus saves the drying time. The time required for the drying is reduced to about $_1/^4$ that of courtyard drying, which is now widely adopted not only in all villages but also in rice mills, which are major small scale industries in India.

Based upon the work done at the Annamalai University, a 10 Ton/day capacity grain drying plant has been erected at Ludhiana, Punjab by the Government of India for crop drying. Another unit of 1/2 Ton per day for cash crop drying such as chillies is erected at Gauhati, Assam, India. A 10 T/day solar paddy dryer is to be erected at Thanjavur, Tamilnadu, India by the Food Corporation of India. These types of dryers use a blower, driven by an electric motor. It is found that we could collect more than 15 times of energy from solar radiation as that used in electric motors. Grain drying particularly after harvest during processing, seed preparation and before storage is a viable project for solar energy application.

Use of Wind Energy, Animal power for driving blowers in a Solar drying unit can eliminate the use of electricity and also make the solar dryers useable in villages where there is no electricity.

MOBILE SOLAR DRIERS FOR AGRICULTURAL APPLICATIONS

In India, we have agriculturists with very small holdings. At present, they carry out the drying operations in the drying yards near their fields adopting the open sun-drying. This has to be changed in order to save food grains, otherwise lost and also save time to market the produce. A portable type of solar drier will be the right type of gadget in Rural India.

Many countries have tried this aspect and in India, works at the experimental level are in progress. The mobile unit can be taken from village to village during the harvest season. Such a unit must be owned by Panchayat, Co-operative Societies, Government Agricultural Departments and other rural societies. The unit must be hired to the villagers and taken to the fields in order to avoid transporting the agricultural produce for the sake of drying purposes. As far as possible, the human and animal power can be used in the solar drying operations in order to avoid use of electricity, which may not be available at the drying yards near the fields in the villages.

The design and development of mobile solar drier will be a challenging problem to Scientists and Engineers if the unit is to be a durable one, withstanding the rough handling and transportation. When once the driers with reasonable cost are made available to the rural public, there will be definitly a great improvement in the agricultural operations and employment potentials in rural India. A portable unit can be utilised for more number of days in a year than a stationary type of solar drier, since the unit can be moved to places where different crops are cultivated during different seasons and hence it will not be idle as in the case of the stationary driers. There are so many gadgets developed for agricultural operations and they are all movable types. Hence the mobile solar drier is a desirable unit that can be added for the rural development.

To conclude, it can be said that low grade thermal energy from simple flat plate collectors can meet the energy needs in villages to a greater extent for drying agricultural produces to preserve food and save fossil fuels.

Solar Energy in Botswana Innovation Centre (RIIC)

by

CHIKUNI, Botswana

Rural Industries Innovation Centre (RIIC) is an organization which is venturing into a number of projects related to solar energy as defined for the purpose of the conference. Among these are 3 types of wind turbine for water pumping and generation of electricity, efficient mud stoves, solar water heaters, solar cookers for domestic and commercial use.

WIND POWER PROJECTS

On the site is a Cretan sail windmill of 2 m diam. on a 12 m tower. The objective was to design a wind energy conversion system at a cost of about P200. The cost of windmill towers are a major factor in the production of wind energy conversion systems. The present tower is of wooden construction to reduce costs. (The possibility of using a wooden tower was demonstrated by the SWD at Twente University of Technology in the Netherlands). Originally a 12 V dc generator was mounted on the rotor. The performance was satisfactory and it was possible to charge accumulators with that system. Our efforts in electricity generation are being hampered because no slow running generator that we know of has been developed. For our purposes, a slow generator would be rated below 300 rpm. If dc, they should purposes, they should preferably be capable of producing a 240 V output. The power output can be anything between 100 W to 1000 W. Lower output dc generators (also slow running) producing 15 W-100 W at 12 V can also be useful to provide electricity for a family (using say low voltage fluorescent lamps). The Cretan sail windmill is presently fitted with a reciprocating transmission to operate a pump. A low cost pump to go with it is presently being developed.

SOLAR ENERGY

Solar cooking

(a) Relatively low cost cookers have been developed. These are simple box type configuration and are ideal for slow cooking. They are completely portable and in fact very much resemble a small suitcase.

Solar Ovens

(b) One type of oven has been quite successful. This is of the reflecting mirror type. Initially a small unit for baking 3-4 loaves of bread was built. Then a large unit was built and is now part of a commercial bakery doing 18 loaves of bread day. A loaf costs 28t (4t cheaper than the other types of bread available) and it is still possible to make a profit of 12t per loaf, i.e. a total profit of Pl. 80 for 15 loaves. This is a considerable income considering that rates in other sectors go down to Pl.00 a day. A solar oven for a family (3-4-loaves a day) costs P45.00. For a commercial bakery costs are around P100.00. There is a lot of interest expressed in the solar ovens and efforts are under way to make this technology more widely known and understood.

Efficient wood burning stoves have been developed from those of Far Eastern Countries. These are made from mud. Although these are quite efficient compared to the open fire ones they have not been found by many to be acceptable as they do not fit into traditional patterns. Dialogue with one village (Moshaneng) resulted in a discussion chart in which points raised for and against each type of stove were recorded. In the end the mud stove had approval from the majority of participants.

Solar Heating

Our basic water heater is made of galvanized pipe sheet and fibre glass and is the most successful and cost effective. They are functional and fairly cheap at P100. Cheaper ones (Pl.00) have also been developed from polyethylene plastic hose and sheet.

Solar Desalination

A solar still is at RIIC for battery water. There is also a test and demonstration model at Ralekgetho Council clinic to desalinate borehole water. This is considered vital as the closest supply is 10 km away.

Photovoltaic Pumping

Moves are underway for RIIC to Co-operate with a German Company for the installation and testing of a 1 kW photovoltaic pumping system in Botswana. This will be an opportunity to asses the solar potential of Botswana. We consider our radiation levels to be quite favourable.

Biogas

Botswana does have a lot of cattle and relies to a great extent on agriculture. Biogas units could produce energy (pumping water and rural lighting). A small unit 0.6 m^3 is operating and is providing an experimental housse with cooking gas and also lighting.

A larger unit 10 m^3 has been constructed. The gas is to be used to run a Methane Engine which has been ordered from India. We are having problems with the gas proofing of the unit. We would be interested in having any information that would lead to simple cheap and effective sealing of the brick (cement plastered) unit and subsquent units.

As one can see we are involved with wide variety of soft technology activities. We will be interested to co-operate with any organizations or firms by way of testing and evaluation of any new relevant technology.

Potential Renewable Energy Applications in Djibouti

by

T.A. LAWAND
Mc Gill University, Canada

Any solar program in the country should envisage applications in the following sectors:

A. Commercio-Industrial, Developed Urban Sector. (This would also apply to developed areas of several towns.)

B. Less Developed Urban Sector.

C. Rural Sector.

DEVELOPED SECTOR

In the first subdivision, it should be noted that there is little industry in Djibouti. The need for industrial heat is limited; the hotels generally don't provide hot water as the water comes out hot from the cold water tap, in summer at least. Nonetheless the use of solar energy for hot water heating for the commercial, residential, institutional (dispensaries and hospitals) and industrial sectors should be given some consideration. The generation of power for the towns and the city of Djibouti is of long term interest to this country. These technologies are not yet mature and certainly, at this stage, too experimental and sophisticated to be considered for use in Djibouti.

The other area of considerable energy usage, such as the transportation sector, the development of renewable energy systems face many years of experimental work before adequate solutions will be found. In any case, Djibouti with its limited technical infrastructure need not be overly concerned with this area at the present time. However, they should keep a watchful eye on all facets of research and development in these fields so they will be in a position to decide when, at some later date, the country should become involved in these areas of technology.

One of the principal electricity consumers in the urban developed sector is that of air conditioning for buildings etc. While some experimental solar powered air conditioning installations have been built in the USA, Japan and elsewhere, these are not yet ready for application in Djibouti. The same could be said about large scale solar refrigeration, which would certainly find application in this region, for cold storage units etc. Practical solar air conditioning is probably hopefully some five or ten years down the road. It may become particularly attractive for high electricity cost areas such as Djibouti.

It should, of course, be noted that while the Ministry of Agriculture is interested in all aspects of renewable energy utilization, it is primarily interested in the applications affecting the rural sectors. It would be difficult to envisage a program in this urban sector lodged under the aegis of the Agricultural Ministry.

194

This however is the crux of the problem; no one in the country seems really seriously concerned with energy or energy costs. After spending one week in Djibouti, the only person who wanted to do something about reducing energy consumption was a hotelier who finds that his electric bills double in the summer owing to the excessive use of air conditioning in the rooms of the hotel. Many systems, like this one, use standard individual window units. They are costly to repair and to operate and to not have a very effective life. Air conditioning is a luxury, albeit a well appreciated one when the temperature soars to 45°C in the summer.

This level of solar energy utilization is relatively costly and sophisticated — far too much in the opinion of this writer, for the means available in Djibouti. This is best illustrated by the comments of the SEDES reports (Ref. 1). It is, in the opinion of the writer, precisely the wrong approach to take for solar energy utilization in this country. Unfortunately, it is the approach that often comes from organizations in industrialized countries, relatively inexperienced in the renewable energy field, but who adopt the classical approach i.e. what have "we" the industrialized society, to "sell" in solar energy technology. The question is almost never posed "can the local population use renewable energy to improve its own situation"? This is not done for two main reasons:

1. It is far more difficult to achieve as it involves introduction training and some commitment to the development of the target area.

2. There is little or no profit in these undertakings.

Hence developing countries will continue to have renewable energies presented to them as though they were some integral off-shoot of the industrial marketplace. Will these exported solar technologies always be adapted or appropriate for the developing area?

In summarizing therefore, there are centainly some applications of renewable energies which can affect the urban — industrial society of Djibouti. A special effort will have to be made to investigate the local energy needs to quantify them and to outline the quantity of this demand. It will then be necessary to determine the cost effectiveness of various renewable energy processes. This has not been done during the Mission due to the lack of time. It is essential, however, that any such program be solidly encompassed in an overall program or policy of energy (and material resources) conservation. Energy conservation effectively doesn't exist in Djibouti, nor is it even thought of in most circles.

LESS DEVELOPED URBAN SECTOR

The bulk of the urban populations of the territory are not the well-to-do, living on imported commodities but local populations with high expenses and earning ordinary salaries. It is in this area that a reduction in the primary energy consumption would be appreciated, not only by the State but also by the consuming population. It is impossible to quantify this energy sector as this would have required some time to assemble and analyse the data. A cursory discussion with E.D.D. officials seemed to indicate that they had not themselves fully analysed this area of energy demand although this data should be easily obtainable as their billing operations are fully computerized. Not all dwellings in the towns and Djibouti City are indeed electrified. However, a quick visit to the homes of some Ministry of Agriculture labourers indicated that a low amperage electricity supply was generally installed.

The principal applications appeared to be for a refrigerator, ceiling and table fans etc. The E.D.D. has a number of specialised tarifs permitting this type of installation.

The principal energy consumer in the homes of the local lower income populations is for cooking fuels. Hence a reasonable program of renewable energy technologies for these areas would appear to be the introduction of:

a) solar cookers and food warmers;
b) improved efficiency wood cooking stoves.

This would reduce the importance of fuel and increase the disposible income of the labouring class. Possibly at a later date, small solar refrigerators or ice makers might also be integrated.

It is doubtful in the short term, that existing solar ice making systems — which are unfortunately themselves, not yet fully developed — could compete with electrically powered refrigerators. However, given the likelyhood of increases in the cost of electric power in the future, these technologies might prove more attractive for application at a later time.

The possibility of using wind powered electric generators as diesel replacement for the outlying cereles may also be given some attention at a later date. This would require an indepth study of the cost of operation of the generation costs of electricity within the country.

In dealing with problems relating to the operation of households within the urban community, it has been suggested that contactable established with the woman's organisations which are active in the country. It may well be that these women can provide the key to the introduction of these technologies in many of these areas. Amongst the more active groups are the Croissant Rouge and the women's section of the Ligue Populaire Africaine pour l'Indépendance.

Combined Electricity and Heat Production

by

Dr. B. HACHEM, Syria

Hybrid systems made up of solar cells for the production of energy, in which the calories given off by the cells are harnessed to meet heating needs, are an attractive prospect in view of the development of complete solar systems for domestic use.

By stressing from now on the importance of these systems, we can only help to further the development of solar energy and make its use more practicable.

We should also note that these combined production systems will have a favourable effect on the cost of photovoltaic energy.

According to our calculations, these hybrid photovoltaic systems perform quite well.

Photovoltaic Concentrators for Water Pumping

by

Dr. B. HACHEM, Syria

I should like to say just this: it is now proven that the use of solar cells under concentration provides an attractive means of water pumping in the developing countries.

The quasi-static functioning of these systems makes them fairly simple to use. The concentration system using a Fresnel linear lens is attractive both technically and economically. The energy produced per unit volume of the concentrator is greater than that of other concentration systems. Combined photovoltaic concentrators would still be economical even if the present costs of solar cells were to be reduced by a factor of 15.

I feel that it would be worthwhile to construct several prototypes of this sort in various parts of the developing countries.

The installation of these prototypes would allow the following five fundamental ideas, proposed for discussion by Chairman Van Overstraeten, to be put into practice:

1. Construction of test fields
2. Exchange of information
3. R & D meetings
4. Training
5. Quality control.

Energy Systems and Design of Communities

by

E.N. CARABATEAS,
National Energy Council of Greece
Ministry of Coordination

Energy systems (production, conversion, recuperation, transfer, distribution, end-use and conservation technologies) enter into the planning of communities in many different ways.

Consequently, the first step in the design of energy systems is a systematic analysis of the relationships existing between these systems and the physical parameters (location, size, type, structure, texture) describing the community.

If the geographic location, size and type of the settlement represent the fixed parameters upon which the entire planning of the community is based, then structure, texture and energy parameters are the variables that must be integrated and optimized. Consequently, the second step in the design of energy systems is to estimate in a very rough way the energy demand of the community on the basis of location, size and type and then manipulate structure, texture and energy variables to fine-tune this demand in accordance with community functions and goals.

As a rule, energy demand varies separately for each end-use over the 24-hour period and during the four seasons of the year. This variation in time must be taken into account since consumption peaks are particularly important in synthesizing a realistic demand profile for planning purposes.

The third step in the design of energy systems is the selection of energy sources to be used and system components to be installed which are best suited to satisfy the projected demand.

Obviously, each end-use can be serviced by a number of alternative system configurations involving different energy source and component combinations. This creates a major problem in that a large number of alternative systems is offered by modern technology for comparative evaluation.

Criteria used to narrow down the available options included:

1. Engineering readiness
2. Efficiency
3. Recuperation potential
4. Scale of commercial applicability
5. Reliability
6. Cost
7. Environmental aspect
8. Integration potential with other energy systems
9. Quality of service to consumers.

Approximate Electric Loads in the Prototype Community

#	Consumers	Type of appliance or consumption	Number of appliances	Power for one appliance or one Consumer (kW)			Daily energy for one appliance or one consumer (kWh)		
				Winter	Spring-Autumn	Summer	Winter	Spring-Autumn	Summer
1	Residential	Refrigerator	900	0.20	0.20	0.20	1.35	1.89	2.70
2	Residential	Deep freezer	45	0.34	0.34	0.34	3.00	3.50	4.00
3	Residential	Washing machine	450	0.50	0.50	0.50	0.25	0.25	0.25
4	Residential	Dishwasher	45	1.00	1.00	1.00	1.00	1.00	1.00
5	Residential	TV sets	900	0.10	0.10	0.10	0.36	0.30	0.30
6	Residential	Lighting	900	0.60	0.50	0.40	1.00	0.70	0.40
7	Residential	Small appliances	864	0.70	0.70	0.70	0.75	0.75	0.75
8	Residential	Cooking	810	2.00	2.00	2.00	3.00	3.00	3.00
9	Schools	Refrigerator	10	0.20	0.20		1.35	1.89	
10	Schools	Lighting		23.28			46.55	40.00	
11	Schools	Other		10.00	10.00		40.00		
12	Health-Welfare Services, Government Services, Town Hall, Banks, Agricultural Cooperative	Refrigerators	23	0.20	0.20	0.20	1.35	1.89	2.70
13	Health-Welfare Services (Lab)	Lighting		0.70	0.70	0.70	4.00	4.00	4.00
14	Health-Welfare Services (Lab)	Lighting		4.00	3.00	3.00	12.00	6.00	18.00
15	Health-Welfare Services (Lab)	Other		3.00	3.00		18.00	18.00	
16	Gendarmerie Station	Lighting		0.25	0.25	0.25	3.50	3.00	2.50
17	Government Services	Lighting		3.50	3.50	2.50	14.00	7.00	12.50
18	Government Services	Other		2.50	2.50	3.00	12.50	12.50	21.00
19	Banks	Lighting		3.00	3.00	3.00	21.00	21.00	21.00
20	Banks	Other		3.00	3.00		21.00	21.00	
21	Agricultural Cooperative	Lighting		1.00	1.00	1.00	4.00	2.00	2.00
22	Agricultural Cooperative	Other		1.00	1.00		2.00	2.00	
23	Town Hall	Lighting		1.50	1.50	1.50	6.00	3.00	3.00
24	Town Hall	Other		1.50	1.50		3.00		
25	Cultural Center	Lighting		3.00	2.00	1.00	12.00	8.00	4.00
26	Cultural Center	Other		2.50	2.50	2.50	8.00	8.00	8.00
27	Agricultural Research Station	Lighting		2.00	2.00	3.00	8.00	4.00	9.00
28	Agricultural Research Station	Other		3.00	3.00		9.00	9.00	
29	Church	Lighting		1.60	1.60	1.60	3.50	3.20	2.90
30	Athletic Center	Lighting		3.00	3.00	3.00	12.00	12.00	12.00
31	Athletic Center	Other		2.00	2.00	2.00	8.00	8.00	8.00
32	Streets	Lighting		180.00	180.00	180.00	1 195.00	956.00	896.00
33	Stores	Lighting		20.00	20.00	20.00	190.00	160.00	130.00
34	Stores	Refrigerator	20	1.50	1.50	1.50	7.50	10.50	15.00

No.	Facility	Appliance							
35	Stores	Freezer	10	1.50	1.50	2.00	10.00	14.00	20.00
36	Stores	Other		20.00	20.00	20.00	160.00	160.00	160.00
37	Village Restaurants	Lighting	15	3.75	3.75	3.75	30.00	22.50	15.00
38	Village Restaurants	Refrigerator		1.00	1.00	1.00	5.00	7.00	10.00
39	Village Restaurants	Freezer	5	1.00	1.50	1.50	7.50	10.50	15.00
40	Village Restaurants	Other		15.00	15.00	15.00	60.00	60.00	60.00
41	Class "B" Hotel	Lighting			20.00	40.00		60.00	90.00
42	Class "B" Hotel	Refrigerator			10.00	14.00		70.00	140.00
43	Class "B" Hotel	Other			20.00	50.00		120.00	200.00
44	Class "C" Guesthouses	Lighting		3.00	5.00	7.50	24.00	30.00	45.00
45	Class "C" Guesthouses	Refrigerator		2.50	3.70	6.50	20.00	36.00	65.00
46	Class "C" Guesthouses	Other		3.50	6.00		28.00	48.00	80.00
47	Camping	Lighting				1.50			7.50
48	Camping	Lighting				7.50			67.50
49	Camping	Refrigerator	50			0.10			1.00
50	Camping	Freezer	1			3.50			35.00
51	Camping	Other	50			0.50			1.50
52	Tourist Restaurants	Lighting Indoor				2.50			12.50
53	Tourist Restaurants	Lighting Outdoor				2.50			12.50
54	Tourist Restaurants	Refrigerator	9			1.00			10.00
55	Tourist Restaurants	Other				10.00			40.00
56	Beach Facilities	Lighting				13.00			39.00
57	Beach Facilities	Refrigerator	5			1.00			10.00
58	Beach Facilities	Other				8.00			39.00
59	Light Industry	Lighting		5.00	3.00		25.00	15.00	136.00
60	Light Industry	Motor drive		36.00	36.00	36.00	136.00	136.00	25.00
61	Light Industry	Other		5.00	5.00	5.00	25.00	25.00	80.00
62	Bakeries	Kneading		10.00	10.00	10.00	80.00	80.00	
63	Olive presses	Motor drive		146.00			3 516.00		
64	Village Restaurants	Cooking		60.00	60.00	60.00	200.00	200.00	200.00
65	Class "B" Hotel	Cooking			45.00	75.00		180.00	300.00
66	Class "C" Guesthouses	Cooking		9.00	18.00	30.00	36.00	72.00	120.00
67	Camping	Cooking				37.50			150.00
68	Tourist Restaurants	Cooking				75.00			300.00
69	Health-Welfare Services	Air-conditioning				14.70			163.30
70	Government Services	Air-conditioning				13.00			86.80
71	Gendarmerie Station	Air-conditioning				2.73			30.36
72	Banks	Air-conditioning				12.60			117.20
73	Agricultural Cooperative	Air-conditioning				6.53			43.60
74	Town Hall	Air-conditioning				7.90			43.60
75	Cultural Center	Air-conditioning				26.00			105.00
76	Agricultural Research Station	Air-conditioning				10.50			65.10
77	Stores	Air-conditioning				107.30			715.00
78	Village Restaurants	Air-conditioning				26.30			190.00
79	Class "B" Hotel	Air-conditioning				98.00			980.00
80	Class "C" Guesthouses	Air-conditioning				29.40			326.00
81		Air-conditioning				7.50			87.50

Twenty-four tasks grouped in four task clusters are required for the design of a total energy system for the community. The first task cluster reviews the data base for the initiation of the design effort. It includes:

1. Review of local conditions (climate and energy resources)

2. Review of state of the art for energy system components

3. Review of state of the art for consumption standards and demand profiles

4. Review of energy conservation techniques.

The second task cluster establishes the type (quality) and scale (quantity) of energy sources available to the community and eliminates from consideration all those considered non-applicable.

The third task cluster estimates the energy demand separately for every sector of consumption (e.g. commercial, residential) and every type of end-use (e.g. lighting, space heating, deep well pumping). Additionally, alternative energy systems for end-use satisfaction are also considered and screened. Based on these data, energy related requirements are developed for use in the physical planning of the community. In return, a community concept plan is produced to allow completion of energy load and demand profile calculations by end-use and by consumption sector. Additionally, sufficient information is generated to help develop an approximate layout of energy distribution networks for use in the design of the energy system.

The fourth task cluster difines energy system alternatives satisfying the projected demand and evaluates their performance characteristics. System optimization is also performed followed by cost and fuel consumption evaluations. Another type of information generated at this stage is the layout and land-use requirements imposed by the various energy systems. Using all this information together with building design data having energy implications, a limited number of detailed community plans is prepared and used to further refine the definition of preferred energy systems.

A basic step towards the design of an energy system for the prototype community is the establishment of demand profiles. This, in turn, requires identification of all expected end-uses (lighting, heating, etc.) for every sector of consumption (residential, commercial, etc.) and specification of the corresponding energy needs (by form and magnitude) created by these end-uses.

Based on an analysis of functional requirements (already established during the physical planning process), the following end-uses were found applicable to prototype community conditions:

A. Thermal Loads

    1. Space heating (residential and commercial)

    2. Water heating (residential and commercial)

    3. Greenhouse space heating

    4. Cooking (residential and industrial)

    5. Bakery oven heating.

202

B. Electric Loads

   6. Deep well pumping for irrigation

   7. Air-conditioning (commercial)

   8. Refrigeration/Freezing (residential and commercial)

   9. Lighting (residential and commercial)

   10. Washing (residential and commercial)

   11. Small appliances (residential)

   12. Radio/TV (residential)

Demand profiles for the various end-uses were then calculated based on established functional requirements and available consumption standards. For electric loads, the demand profiles were computed for three different seasonal periods (winter, spring-autumn and summer) and for two different typical days (workdays and holidays). A detailed breakdown of electric power consumption by end-use category (Table 1 was coupled with standard demands to provide the primary data input for computer-based calculations. In all, 500 different demand profiles were generated allowing synthesis of typical total daily load curves.

Table 2 shows the daily electric demands separately for each of the three seasonal periods mentioned above.

TABLE 2

Total electric demand profiles

Seasonal period	Maximum daily consumption (kWh)	Average daily peak demand (kW at consumers' station)	Maximum daily peak demand (kW at consumers' station)	Required power of prime mover (kW)
Winter	12,175	685	822	970
Spring Autumn	15,680	730	865	990
Summer	19,340	864	950	1 120

Alternative energy sources which could be used single or in combination to satisfy these community demands were narrowed down to the following;

1. Biomass energy

2. Solar energy

3. Wind energy

4. Ocean surface thermal energy

5. Energy from fossil fuels (solid, liquid and gaseous)

6. Energy from the local Electricity Network.

Biomass production can reach 6,810 tons/year, a value equivalent to 2,757 tons of oil.

Solar radiation can yield a maximum of 1,940 kWh/m² per year for a non-tracking collector fixed at 30° tilt angle. For a full tracking collector the maximum energy increases to 2,516 kWh/m².

The amount of useful energy that can be extracted per m² of turbine blade area is 423 and 317 kWh/year for the horizontally and the vertically mounted rotor (Darrieus) types respectively. The theoretically maximum energy that can be extracted is 369 kWh/m²-year. Wind energy can be used either for supplementing other energy sources or for meeting small independent energy needs. Note in this respect that the largest commercial units for the conversion of wind energy to mechanical power are currently limited in the range of 14 kW.

The availability of year-round warm ocean water offers a thermal source that can be exploited through the use of heat pumps. This scheme, however, presupposes that the prototype community is located on the water front.

Fossil fuels (not locally available) require transportation from areas outside of Crete, or even Greece, and this extra task adds to their cost. Note that the cost of transportation becomes increasingly important as the thermal capacity per unit weight or volume of fuel drops.

Electricity can be supplied directly to the community from the local Electricity Network which operates at 15 kV with a 1,300 kVA carrying capacity, enough to satisfy present and near future energy needs.

Application of the procedural methodology to the design of the total energy system for use by the prototype community revealed that different technologies based on different energy sources can be employed for servicing each end-use. A common denominator, however, tying all these applicable technologies together is the power generating system viewed both as an electric and thermal energy source.

System combinations for power generation that are best suited for exploiting energy sources available to the prototype community, that are compatible with the anticipated community size and that have same order of magnitude costs, include:

1. Diesel (with jacket water and exhaust gas heat recovery).

2. Diesel-Biomass (combustion products provide thermal energy to the working fluid of an Organic Rankine Cycle).

3. Biomass-Solar (parabolic, line-focusing, single-axis tracking collectors requiring approximately 10⁵ m² of land surface.

4. Tie-into the local Electricity Network.

5. All Solar (this system is included for comparison purposes only, since cost and sunshine exposure data mitigate against this approach).

Supplementary systems for driving non-electric (thermal) loads must also be selected from among alternatives and to the degree possible must be integrated with each of the electric power generating schemes.

Obviously, however, texture parameters, entering the physical planning of the community, are expected to affect the design (and, therefore, cost) of the energy systems. Indeed, texture plays a role in two important ways:

1. Through differences in thermal losses from buildings caused by site-specific variations of their exterior surfaces and their degree of compactness.
2. Through differences in thermal losses and construction costs of energy supply networks caused by variations in network length.

Consequently, alternative energy systems must be adapted to selected community plans, if results are to retain their full measure of usefulness. Two alternative community plans, plan "B" and plan "A", have been selected for this purpose. The difference between these two plans lies in the density of community structures.

Greenhouse space heating and bakery operation are serviced through direct biomass combustion. Associated costs are independent of community plan layout.

Energy Farming in Coastal Deserts

by

M. WAGENER, Federal Republic of Germany

Talking about biomass as energy carrier in developing countries we have to keep in mind that substantial parts of the southern hemisphere are deserts with no or low productivity. A chance have coastal deserts, because there is at least seawater available.

A system of energy farming using marine algae grown in shallow ponds is presently under investigation, financed by the EC. It has favorable economic aspects and, in addition, could give a basis for industrial development for the poor areas.

Underground Water in the Service of Development

by

L. MONITION, France [1])

Water is a critical factor in urban, agricultural and industrial development throughout the world, but methods of water exploitation and management vary according to whether we are living in arid zones or more temperate climates. Information on underground water, however, is the basic data for the assessment of overall water resources, and it is the task of the hydrogeologist to provide planners and decision-makers with clear and precise information on the subject.

In response to a request from the Fonds d'Aide et de Coopération de la République Française [2]) (FAC) and the Comité Inter-Africain d'Etudes Hydrauliques [3]) (CIEH) of Ouagadougou (Upper Volta), the Bureau de Recherches Géologiques et Minières [4]) (BRGM) prepared in the years 1975-76 a series of documents concerned solely with underground water, which could be utilized directly in the regional or national planning of the **Sahel countries:** Mauretania, Senegal, (Gambia), Mali, Upper Volta, Niger, (Cameroon) and Chad.

These cartographic documents, on the scale 1/1 500 000, relate to four planning factors:

— the productivity of water-bearing beds and the initial delivery rates of the installations (wells or bore-holes);

— the cost of water;

— water quality and the suitability of the water for irrigation;

— water ressources.

The thirty known aquifers have been classified as **generalized aquifers** and **discontinuous aquifers,** these terms being applied to underground reservoirs of different characteristics, which can be exploited at varying rates depending on local conditions.

1. Chart of the productivity of water-bearing beds and the initial delivery rates of installations

This chart shows, for homogeneous zones, the probable daily delivery rate of a bore-hole sunk to an aquifer; the trend of delivery over time is disregarded. It is assumed that depletion should not exceed one third of the depth of an unconfined replenishable aquifer, and should not be more than 100 m (where this is possible)

---

[1]) BRGM, Orléans, France.
[2]) Aid and Cooperation Fund of the French Republic.
[3]) Inter-African Committee on Hydraulic Studies.
[4]) Geological and Mining Research Bureau.

in the case of confined aquifers. The productivities evaluated range from 50 m³/day to over 2200 m³/day for generalized aquifers, and from less than 25 m³/day to 75 m³/day for discontinuous aquifers.

2. Chart of the average cost of exploiting underground water

This chart shows the average investment and operating costs per cubic metre of water in 1974 CFA francs, taking account of the productivity data provided in the chart referred to above. The parameters taken into account are the following: cost of bore-holes, cost of pumping equipment and cost of engine; amortization over 15 years for the bore-holes, five years for the pump and four years for the engine; also the fuel, oil and grease consumption of the engines and associated costs. The cost has been assessed per m³ carried to a height of 10 m above the ground on the basis of 3200 hours of operation a year. The costs range from less than 10 CFA francs (1974) to over 50 CFA francs. This calculation could not be performed for the discontinuous aquifers.

3. Chart of the quality of water for irrigation purposes

This chart was prepared in accordance with the Wilcox norms on the basis of the resistivity of the water and the rate of sodium absorption

$$(SAR = \frac{Na}{\sqrt{^1/_2\ Ca + Mg}}).$$

Five grades were distinguished ranging from "excellent" to "poor".

4. Chart of resources

This chart complements and summarizes the three charts already described. It specifies the nature of the aquifers (generalized or discontinuous), the depth of the sub-surface water (isobathic curves), the chemical quality expressed as total concentration, and the usable resources.

Among the usable resources, a distinction must be made between resources that are **renewable** from precipitation expressed in m³/year/km² and **exploitable** resources, i.e. the portion of the reserves that can be exploited under certain technical and economic conditions, knowing that the quantity withdrawn will not be replaced. This type of exploitation can be compared to the exploitation of hydrocarbon depostis. In unconfined aquifers the water level may be lowered over the whole aquifer by one third of its depth, but not beyond 100 m below ground level; for confined aquifers, the level has been fixed at 100 m below the ground surface.

Under these conditions, the **renewable resources** over the area of 4 million km² in question be 60 km³/year, and the **exploitable reserves** would be from 800 to 1800 km³. It is these latter quantities that ought to be more effectively exploited in order to facilitate the economic "take-off" of certain regions of the Sahel.

Evaluation of Renewable Resources and Exploitable Reserves

Country and total area in km²	Area of the part of the country covered by the chart in km² and % of the total area	Renewable resources in km³/year	Exploitable reserves in km³
Cameroon (475 442)	96 000 (20%)	5,4	11 to 23
Gambia (10 500)	10 500 (100%)	1	9 to 19
Upper Volta (274 200)	272 400 (99%)	6,2	1 to 3
Mali (1 201 625)	807 000 (67%)	13,1	82 to 186
Mauritania (1 030 700)	471 000 (45%)	0,2	54 to 119
Niger (1 267 000)	988.900 (78%)	4,6	326 to 672
Senegal (195 500)	195 500 (100%)	9,4	86 to 184
Chad (1 284 000)	1 049 200 (82%)	20,7	260 to 538
TOTAL	3 890 500	60,6	829 to 1 744
FRANCE	550 000	100 to 120	several hundred

¹) 1 km³ = 1 000 million m³.

209

D. LIST OF PARTICIPANTS

ALGERIA	Mr. A. BOUCHOUAREB Soc. Nat. d'Electricité et du Gaz 2, Bouakouir Alger
	Dr. N. DJEGHRI SONATRAC 10, rue du Sahara, Alger
	Mr. A. HAMROUR SONATRACH 10, rue du Sahara, Alger
	Mr. M. A. LARBI Ministère de l'Energie et des Industries Pé- trochimiques Alger
	Mr. B. OULD HENIA O.N.R.S. 27, Abri Arezhi Hydea, Alger
ARGENTINA	Dr. M.B.A. CRESPI Comisión Nac. de Energía Atómica Avda. del Libertador 8250 1425 Buenos Aires
	Ing. J.A. LEGISA Fundación Bariloche Casilla de Correo 138 8400 Bariloche
	Dr. L.R. SARAVIA Presidente Asociación Latino Americana de Energia Solar Universidad Nacional Salta
BANGLADESH	Dr. A. ASGAR Associate Professor, Bangladesh University of Engineering & Tech- nology Dacca
	Dr. S.H. CHOUDHURY Agricultural University Mymensingh

BARBADOS	Dr. J. TUDOR Head - Laboratory Section Barbados National Standards Institution "Flodden" Culloden Road St Michael
BELGIUM	Prof. J. BOUGARD Faculté Polytechnique de Mons Dept. de Chimie 9, rue de Houssain 7000 Mons
	Mr. F. BRAIBANT Cockerill 4100 Seraing
	Mr. R. CHOMÉ Association Internationale de Développement Rural 20, rue du Commerce 1040 Brussels
	Prof. P.J.A. de MEESTER Katholieke Universiteit Leuven De Croylaan 2 3030 Leuven
	Mr. M. OSTERRIETH Service de la Programmation de la Politique Scientifique 8, rue de la Science 1040 Brussels
	Mr. M. SHUBAN Journaliste Grotstraat 49 1900 Overijse
	Mr. A. STENMANS Secrétaire Général des Services du Premier Ministre pour la Programmation de la Politique Scientifique 8, rue de la Science 1040 Brussels
	Prof. R. VAN OVERSTRAETEN Dept. Electrotechniek Katholieke Universiteit Leuven Kardinal Mercierlaan 94 3030 Heverlee

BENIN	Mr. E. PARAISO Directeur Général de la Société Béninoise de Distribution d'Eau et d'Electricité Cotonou
	Mr. J. GNIDEHOU Directeur de la Planification Ministère du Plan Cotonou
BOLIVIA	Prof. C.B. AGUIRRE Instituto de Energia Academy of Sciences Casilla 5279 La Paz
	Prof. A. TREPP Director Instituto de Energia Universidad San Adres Casilla 1923 La Paz
BOTSWANA	Mr. E. CHICUNI Rural Industries, Innovation Centre, Box 138 Kanye
BRAZIL	Eng. M. de Paula FERNANDES Coordinator of Energy Programme Fundaçâo Centro Tecnologico de Minas Gerais (CETEC) Belo Horizonte
	Dr. J.M. PINTO Ministério das Minas e Energia Esplanada dos Ministérios - Bloco J 70056 Brasilia (DF)
	Dr. C. TRINDADE Executive Director Centro de Tecnologia Promon, CTP Praia do Flamengo 154 22210 Rio de Janeiro
BURUNDI	Mr. L. KAGISYE Chef de Division, Mines et Energie Ministère du Plan B.P. 2222 Bujumbura

Mr. P. NIYIMBONA
Directeur du Dépt. de l'Energie
Gouvernement du Burundi
B.P. 1860
Bujumbura

CAMEROON

Mr. A. SIMO
Office National de Recherches Scientifiques et
Techniques ONAREST
Yaounde

Mr. J.P. TSANGUE
Régie Nationale des Chemins de fer
B.P. 304
Douala

CANADA

Dr. J. HOLLINS
Conservation and Renewable Energy Branch
Dept. of Energy, Mines and Resources
580 Booth Street
Ottawa KlA OE4

Mr. R. WILSON
Policy Branch, Canadian International Development Agency
Rue Principale 200
Hull

CAP-VERT

Mr. M.F. QUERIDO
Directeur Commission Technologie
B.P. 30
Praia

CHILE

Prof. G. FRICK
Laboratorio de Termodinamica
Universidad Catolica de Valparaiso
Alvarez 454-2
Vina del Mar

Dr. Ing. P. ROTH
Universidad Santa Maria
Cas. 110-V
Valparaiso

CONGO

Mr. A. NTOUMI
Société Nationale d'Electricité
Brazzaville

COSTA RICA

Dr. E. GONGORA
Casa Presidencial
San José
Costa Rica

DENMARK	Dr. P. NORDGAARD HANSEN Laboratoriet for Varmeisolering Danmarks Tekniske Højskole Bygning 118 2800 Lyngby
	Prof. P. WORSØE-SCHMIDT Technical University of Denmark 2800 Lyngby
DJIBOUTI	Mr. A. ANIS Directeur de l'Institut Supérieur pour Etudes et Recherches Scientifiques Djibouti
ECUADOR	Mr. R. MALDONADO National Institute of Energy Santa Prisca 223 Quito
	Mr. E. TORO El Comercio Casilla 2698 Quito
EMPIRE CENTRAFRICAIN	Dr. J.-M. BASSIA Assistant à la Faculté des Sciences de Bangui Bangui
ETHIOPIA	Mr. A. A. TILAHUN Permanent Secretary Ministry of Mines, Energy and Water Resources Addis Ababa
	Dr. Ing. G. WOLDEGHIORGHIS Executive Secretary of Ethiopian National Energy Commission P.O. Box 486 Addis Ababa
FRANCE	Mr. R. BAZIN Pompes Guinard 172, Bd. St. Denis 92402 Courbevoie
	Mr. M. BRUNET La Radiotechnique Compelec RTC 130, Av. Ledru Rollin 75540 Paris Cedex 11

Mr. M. BULLIO
C.I.A.R.D.
Comité Français des Concours d'Inventions et
Innovations adaptées aux Régions en Développement
42, rue Cambronne
75015 Paris

Dr. P. CHARTIER
Institut National de la Recherche Agronomique
INRA
Route de Saint Cyr
78000 Versailles

Mr. CHASSERIAUX
Ministère de l'Industrie
101, rue de Grenelle
75007 Paris

Mr. J. COUTURE
Président de la Commission de l'Energie du
7ᵉ Plan
Société Générale
29, Bd Haussemann
75009 Paris

Mr. J. DAMAGNEZ
Station de Bioclimatologie, INRA
84140 Montfavet

Mr. B. DEVIN
Centre d'Etudes Nucléaires de Saclay (CEA)
B.P. No. 2
91190 Gif-sur-Yvette

Dr. H. DURAND
Président
Commissariat à l'Energie Solaire
208, rue Raymond Losserand
75014 Paris

Mr. R. GICQUEL
Ecole des Mines
60, Bd. St. Michel
75272 Paris

Mr. J.P. GIRARDIER
SOFRETES
B.P. 163
45203 Montargis

Mr. GOURIOU
Bureau Y. HOUSSIN
65, Av. Marceau
75116 Paris

Mr. P. JOURDE
CEA - DAM, Direction des Essais
B.P. 561
92542 Montrouge

Mr. M. LESSIEUR
Compagnie Française des Pétroles C.F.P.
5, rue Michel Ange
75016 Paris

Mr. M. LUCAS
CNEEMA
Parc de Tourvoie
92160 Antony

Mr. MERCIER
Leroy Somer
Angoulême

Mr. M. MERIGOUX
Alsthom-Atlantique
141, rue Rateau
93123 La Courneuve

Mr. M. MEUNIER
S.E.M.A.
16-18, rue Barbès
92128 Montrouge

Mr. L. MONITON
Bureau de Recherches Géologiques et Minières,
B.R.G.M.
B.P. 6009
45018 Orléans-Cédex

Mr. J. PHELINE
Commissariat à l'Energie Atomique (CEA)
29-33, rue de la Fédération
75015 Paris

Mr. A. G. PLUCHARD
CNET
39-40, rue du Général Leclerc
92131 Issy les Moulineaux

Mr. M. RODOT
C.N.R.S.
15, Quai Anatole France
75007 Paris

Mr. B. SARRE
France Photon
B.P. 119
1600 Angoulême

Mr. G. TOTH
Commissariat à l'Energie Solaire
208, rue Raymond-Losserand
75014 Paris

Mr. R.L. VIC
Compagnie Générale d'Electricité
Laboratoire de Marcoussis
Route de Nozay
91460 Marcoussis

GABON

Mr. S. GASSITA
Directeur Général de l'Energie et des Ressources
Hydrauliques
B.P. 874
Libreville

GAMBIA

Mr. M. SAHO
Meteorological Officer
Dept. of Hydrometeorology
Banjul

GERMANY (F.R.)

Prof. W.H. BLOSS
Institut für Physikalische Elektronik
Universität Stuttgart
Böblinger Str. 70
7 Stuttgart 1

Dr. R. BRIECHLE
AEG
Elisabethenstr. 3
7900 Ulm

Mr. V. CORDES
AEG
Hasenkamp 1
2000 Wedel

Dr. FRANCK
Kraftwerk-Union
852 Erlangen

Mr. F. FRIEDRICH
Kernforschungsanlage Jülich
Postfach 1913
5170 Jülich

Dr. J.W. GRÜTER
Kernforschungsanlage Jülich
Postfach 1913
5170 Jülich

Mr. P.J. HEINZELMANN
Bundesministerium für Forschung und Technologie
Stresemannstr. 2
5300 Bonn Bad Godesberg

Dr. HENSELER
Dornier
Postfach 1360
7990 Friedrichshafen

Dr. G. ISENBERG
M.A.N., Neue Technologie
Dachauerstr. 667
8000 München 50

Dr. F. JÄGER
Institut für Systemtechnik und Innovationsforschung
Fraunhofer-Gesellschaft
Sebastian-Kneipp Str. 12-14
7500 Karlsruhe 1

Prof. G. LEHNER
Universität Stuttgart
Institut für Theorie der Elektrotechnik
Breitscheidstr. 3
7000 Stuttgart

Mr. LEWERENZ
Bundesministerium für Wirtschaftliche Zusammenarbeit
Karl Marx Str. 4-6
5300 Bonn 1

Dr. LINNEBORN
Imbert Energietechnik
Steinweg 11
5760 Arnsberg 2

Dr. M. MELISS
Kernforschungsanlage Jülich
Postfach 1913
5170 Jülich 1

Dr. H.A. COMATIA
Institut für Energie- und Kraftwerkstechnik an der Universität Essen
4300 Essen

Dr. SCHMIDT-KÜNTZEL
Bundesministerium für Forschung und Technologie
Heinemannstr. 2
5300 Bonn Bad Godesberg

Dr. K.P. SCHUBERT
Dornier
Postfach 1360
7990 Friedrichshafen

Dr. A. STREHLER
Landtechnik Weihenstephan
Vöttinger Str. 36
8050 Freising

Dr. H. TREIBER
MBB
Ottobrunn, Postfach 80 11 69
8000 München

Prof. K. WAGENER
Kernforschungsanlage Jülich
Postfach 1913
5170 Jülich

Prof. F. WIENEKE
Universität Göttingen
Gutenbergstr. 33
Göttingen

GHANA

Dr. I. OKOH
Forest Products Research Institute
University Box 63
Kumasi

Prof. F.O. KWAMI
U.S.T. Kumasi
University Post Office
Kumasi

GREECE

Mr. E.N. CARABATEAS
National Energy Council
Ministry of Coordination
Stadiou 3
Athens

Messrs. E. CONSTANTINIDES and G. KON-
TOROUPIS
Society of Mechnical Engineers of Greece
Athens

Mr. COTZAMBASSI
ELOT
Zacynthou 42
Athens

Mr. M. DIMOS and I. NADUM
Technometal
Panepistimiou 59
Athens

Dr. M. MELISSAROPOULOU (Mrs.)
Solar Energy Program
Scientific Research and Technology Service
Ministry of Coordination
Athens

Mr. C. TSILALIS
Société Architect
Frantzi 10 T.T.
708 Athens

Mr. A. VOYATJIS
Elot Architect
Haritos 8
Athens

GUINEA

Mr. K.K. KOUROUMA
Directeur des Sciences Exactes à l'Institut
National des Recherches et Documentation
(INR/DG)
B.P. 561
Conakry

GUINEA BISSAU

Mr. F. VAZ MARTINS
Commissaire d'Etat à l'Energie, l'Industrie et les
Ressources Naturelles
C.P. 353
Bissau

GUYANA

Mr. B. GIBBS
Senior Officer for Science and Industry
National Science Research Council
P.O.B. 689
Georgetown

INDIA

Dr. T.K. BHATTACHARYA
Project Manager
Central Electronics Ltd.
Site 4, Industrial Area
Sahibabad 201005 U.P.

Dr. C.L. GUPTA
Tata Energy Research Institute
Sri Aurobindo International
Centre of Education
Pondicherry 605002

Prof. V.R. MUTHUVEERAPPAN
Annamalai University
K.R.M.C. Avenue
Annamalainagar 608101

Dr. H.N. SHARAN, Director
Bharat Heavy Electricals Ltd.
18-20 Kasturba, Gandhi Marg
New Delhi 110001

Dr. R.K. SURI
Bharat Heavy Electricals Ltd.
Pusa Rd 605
Vikram Tower
Rajendra Place
New Delhi 110008

INDONESIA

Mr. S.H. NASUTION
Jakarta Pusa Indonesia
Jakarta

Mr. SASTROAMIDJOJO
Faculty of Science
Gadjah Mada University Sekip III
Yogyakarta

IRELAND

Dr. M. NEENAN
Agricultural Institute
Oakpark
Carlow

Mr. T.C. OBRAIN
Industrial Development
Authority
Langsdown House
Dublin 16

Mr. D. O'DOHERTY
National Board for Science and Technology
Shelbourne House
21, Shelbourne Road
Dublin 4

ISRAEL

Mr. L. BRONICKI
Ormat Turbines Ltd.
P.O. Box 68
Yavne

223

Mr. J.L. LANDO
"Tadiran" Israel Electronics
P.O. Box 648
Tel Aviv

Mr. A. SHAVIT
Head of Research & Development Division
Ministry of Energy and Infrastructure
University str.
Tel Aviv

Dr. H. TABOR
The Scientific Res. Foundation
P.O. Box 3745
Jerusalem

ITALY

Mr. A.M. ANGELINI, Director General,
ENEL — Ente Nazionale per l'Energia Elettrica
Via G.B. Martini 3
00198 Rome

Mr. G. ANGELUCCI
ELC Electroconsult
Via Chiabrera 8
Milan

Dr. ANTONIOZZI
Minister for the Coordination of Research
Rome

Mr. G. BEER
Phoebus
Viale Leopardi 148
Catania

Mr. G. BENEVOLO
AGIP S.p.A., (ENI Group)
Via E. Mattei
Rome

Mr. F. BERNARDI
Sogesta S.p.A.
Via Crucicchio
Urbino (PS)

Dr. G.D. CANEPA
Direzione Sviluppo, ALITALIA
Palazzo Alitalia
Piazzale Pastore
Rome

Ing. G. CASTELLI
ENEL
Via G.B. Martini 3
00198 Rome

Ing. E. COLOMBO
Instituto Economia delle Fonti di Energià
Università Bocconi
Milan

Dr. C. CORVI
ENEL
Via G.B. Martini 3
00198 Rome

Dr. CUOZZO
C.T.I.P. Solar S.p.A.
22 via Po
00198 Rome

Mr. A. COCCHI
Università di Bologna
Viale Risorgimento 2
Bologna

Mr. N. DE BENEDETTI
TTS SPAZ
Via G.B. Morgagni 50
Rome

Prof. P. DELLA PORTA
Vice Presidente Consigliere Delegato
S.A.E.S. Getters
Via Gallarate 215
Milan

Mr. E. DOMINCO
ELC Electroconsult
Via Chiabiera 8
Milan

Prof. G. ELIAS
Consiglio Nazionale delle Richerche
Via Morgagni
Rome

Dr. G. FIORITO
Centro Studi Energia
Via Serra 6
Genova

Prof. F. FITTIPALDI
Instituto di Fisica
Facoltà di Ingegneria dell'Università
Piazzale Tecchio
Naples

Dr. R. FLORIS
Ansaldo S.p.A.
N. Lorenzi 8
16152 Genova

Mr. FOLCHI
Ministry of Industry
Rome

Mr. V. GUIDA
SNAM Progetti
Viale de Gasperi S. Donato
Milan

Mr. P. LICATA
ENI
Via E. Mattei
Rome

Mr. MARASCHINI
Regione Lombardia
Milan

Mr. C. MICHELI
SNAM Progetti
Viale de Gasperi S. Donato
Milan

Prof. R. MORANDINI
Istituto Sperimentale per la Selvicoltura
Viale Santa Margherita 80
Arezzo

Prof. PANATI
Instituto Economia Fonti Energia
Università Bocconi
Milan

Prof. S. PIZZINI
Istituto G. Donegani
Montedison
Largo Donegani 1/2
Milan

Ing. QUILICO
FIAT (SES — Settore Energia)
Via Cuneo 20
Torino

Dr. R. REVELLI
Società Metallurgica Italiana
Borgo Pinti 99
Firenze

Messrs. ROCCA and SAPORITI
Ministero Beni Culturali
Rome

Ing. A. ROSSI
SNAM Progetti
Viale de Gasperi S. Donato
Milan

Dr. M. DE SANTI
Gruppo Tecnico Problemi Energetici
Regione Toscana
Firenze

Dr. P. SEMIANI
ENI
Piazza Mattei
Rome EUR

Ing. G. SIMONI
Direzione Generale Fonti Energia e Industrie di
Base
Ministero Industria
Rome

Ing. A. TASCHINI
ENEL
Bastioni Porta Volta 10
Milan

Ing. P. VENDITTI
Comitato Nazionale per l'Energia Nucleare
Viale Regina Margherita 125
Rome

IVORY COAST

Mr. D. KROKO
Directeur des Etudes et de la Recherche Tech-
nologique, Energie Electrique de la Côte-d'Ivoire
EECI
B.P. 1345
Abidjan

Mr. T. MANOU
Dept. de l'Aménagement et de l'Habitat Rural,
Office National de Promotion Rurale (ONPR)
B.P. V 165
Abidjan

JAMAICA

Dr. K.C. LEE (Mrs.)
Scientific Research Council
P.O. Box 350
Kingston 6

JORDAN	Dr. F.A. DAGHESTANI Deputy Director General Royal Scientific Society P.O. Box 6945 Amman
	Mr. A.Y. ENSOUR Director General Jordan Electricity Authority P.O. Box 2310 Amman
	Mr. S. KAWAR Director of Planning Telecommunications Corporation P.O. Box 1689 Amman
	Prof. S. QASEM, Dean, Faculty of Agriculture University of Jordan Amman
KENYA	Dr. P.M. GITHINJI Institute of Mechanical Engineering University of Nairobi P.O. Box 30197 Nairobi
	Mr. MWHIHIA Deputy Secretary University of Nairobi P.O. Box 30197 Nairobi
LESOTHO	Mr. E.K. MOTOPI Architect Ministry of Works Maseru
LIBERIA	Mr. L. KELLER Planning Manager Liberia Electricity Corp. Liberia
MADAGASCAR	Mr. J.B. RAMANANA Service de Physique Etablissement d'Enseignement Supérieur B.P. 906 Université de Madagascar Antananarivo

Mr. E. RABEDAORO
L.E.R.
Chef Département Etudes-Programmes-
Laboratoires, SOLIMA
B.P. 433
Tamatave

MALAWI

Mr. E.C. LAWSON
Head of Department of Agricultural Engineering
BUNDA College of Agriculture
University of Malawi
Lilongwe

MALAYSIA

Dr. G.S. CHUAH
School of Physics
Universiti Sains Malaysia
Minden
Penang

Mr. N.J. MONERASINGHE
University Science Malaysia
Minden
Penang

Mr. M.B. YUSOFF
Research Officer
Standards and Industrial
Research Institute of Malaysia
P.O. Box 35
Shah Alam
Selangor

MALI

Mr. A. SIDIBE
Ingénieur
Laboratoire National de l'Energie Solaire
Bamako

Mr. B. VERSPIREN
Ingénieur agronome
Mali Aqua Viva
B.P. 1
San

MAURITANIA

Mr. Y. DIALLO AHMED
Chef de la Division Etudes et Travaux,
SONADER
Nouakchott

Mr. M. KANE
SONELEC
B.P. 355
Nouakchott

MAURITIUS

Mr. S.C. HOW PAK HING
Lecturer in Mechanical Engineering
University of Mauritius
Reduit

Mr. J.M. PATURAU
Chairman
Joint Economic Committee
Place d'Armes
Port-Louis

MEXICO

Dr. J.I. HERRERA
Dirección General de Aprovechiamiento de las
Aguas Salinas y Energia Solar
Boulevard Pípila 1
Presa San Joaquin, Tec
Mexico 10 D.F.

Mr. R. NANJARREZ
Dirección General de Approvechiamiento de
Aguas Salinas y Energia Solar
Boulevard Pípila 1
Presa San Joaquin, Tec
Mexico 10 D.F.

Prof. E.J. PEREZ, PhD
Centro de Investigación del Instituto Politécnico
Nac.
Apartado Postal 14-740
Mexico D.F.

Mr. SAMUEL
Dir. Gen. de Approvechiamiento de Aguas
Salinas y Energia Solar
Boulevard Pípila 1
Presa San Joaquin, Tec
Mexico 10 D.F.

THE NETHERLANDS

Ir. C. DEN OUDEN
Institute of Applied Physics
P.O. Box 155
Stieltjesweg 1
Delft

Dr. P. SMULDERS
Eindhoven University of Technology
P.O. Box 513
Eindhoven

Dr. L. VERMIJ
Technical Manager
HOLEC Switchgear Group
P.O. Box 23
Hengelo (OV)

NIGER

Dr. A. MOUMOUNI
Directeur de l'ONERSOL
B.P. 621
Niamey

Mr. A. SEINI
Directeur du Génie Rural
Ministère du Développement Rural
B.P. 241
Niamey

NIGERIA

Mr. C. EZEILO
University of Nigeria
Nsukka

NORWAY

Mr. T.H. DIGERNES
Christian Michelsen Institute
Fantoftvegen 38
5036 Fantoft-Bergen

PAKISTAN

Dr. M.M. QURASHI
Appropriate Technology Development Organi-
sation
Islamabad

Mr. N.H. QURESHI
Director General
Energy Resources Cell
Ministry of Petroleum and Natural Resources
Islamabad

PAPUA NEW GUINEA

Dr. Q.A. AHMAD
Senior Lecturer for Engineering
University of Technology
P.O. Box 793
Lae

PERU

Dr. M. DURAN
Math. and Physics Department
Universidad Huamanga
Ayacucho

PHILIPPINES	Mr. R.P. VENTURINA National Science Development Board Gen. Santos Bicutan Tagig, Metro Manila
RWANDA	Dr. F. KALOS Directeur du C.E.A.E.R. Université Nationale du Rwanda B.P. 117 Butaré
	Mr. C. NTAKIRUTINKA C.E.A.E.R. Université Nationale du Rwanda B.P. 117 Butaré
SENEGAL	Mr. O. DIALLO Maître Assistant de Physique Faculté des Sciences de Dakar Dakar
	Prof. D.F. FALL Directeur de l'Institut de Physique Météorologique H. Masson Université de Dakar B.P. 476 Dakar
	Mr. S. LATYR Institut de Physique Météorologique H. Masson Université de Dakar B.P. 476 Dakar
	Mr. G. MADON Ingénieur Institut de Physique Météorologique H. Masson Université de Dakar B.P. 476 Dakar
SIERRA LEONE	Dr. M.W. BASSEY Faculty of Engineering Fourah Bay College University of Sierra Leone Freetown
SINGAPORE	Mr. Thong-Ngee GOH Faculty of Engineering University of Singapore Singapore 5

SOMALIA	Mr. A. ISAQ Direttore Progetto "KUDHA" Ministero Pianificazione Mogadiscio
SPAIN	Mr. V. VALVERDE Ingeniero Industrial Centro de Estudios de la Energía Agustín de Foxá, 29 Madrid
SRI LANKA	Dr. S. GNANALINGAM Head, Section of Applied Physics and Electronics Ceylon Institute of Scientific and Industrial Research P.O. Box 787 Colombo
SUDAN	Prof. A.A.R. ELAGIB Chairman, Scientific and Technological Council for Research P.O. Box 6094 Khartoum
SWEDEN	Mr. S. FAUGERT Head of Section, Ministry of Industry Fack S-103 10 Stockholm Mr. B. SWEDEMAR Head of Section Board of Technical Development Fack 10072 Stockholm
SWITZERLAND	Dr. P. KESSELRING Eidg. Institut für Reaktorforschung 5303 Würmlingen Mr. J.R. MÜLLER Institut de Thermique Appliquée Ecole Polytechnique Fédérale Halle de Mécanique Ecublens 1015 Lausanne
SYRIA	Dr. BECHER HACHEM C.E.R.S. P.O. Box 4470 Damascus

TANZANIA	Mr. M.A. GUJARATI Lecturer in Mechanical Engineering University of Dar Es-Salaam Sinza Road Dar Es-Salaam
	Mr. A.A. LYAMCHAI Technologist Small industries development organisation P.O. Box 2476 Dar Es-Salaam
THAILAND	Mr. P. SUCHINDA Director of Rural Development Project Division National Economic and Social Development Board Krungkasem Road 962 Bangkok
	Mr. T. WETCHAKARUN Electricity Generating Authority Charunsanitwong Monthaburi
TOGO	Mr. G. MENSAH Chef du Bureau d'Etudes du CCL B.P. 1762 Lomé
	Mr. A.G. OSSENI Direction de l'Hydraulique et de l'Energie B.P. 335 Lomé
TRINIDAD & TOBAGO	Prof. S. SATCUNANATHAN University of West Indies Faculty of Engineering 65, Cheeseman Avenue St. Augustine 6624752
TUNESIA	Mr. A. KOSSENTINI Faculté des Sciences et Techniques SFAX Tunis
	Mr. T. BSILA La Steg Tunis
	Mr. MILI La Steg Tunis

UNITED KINGDOM

Dr. M.D. ASHTON
Department of Energy
Thames House South
Millbank
London SW1P 4QJ

Prof. P.D. DUNN
Department of Engineering
University of Reading
Whiteknights
Reading Berks RG6 2AY

Prof. D.O. HALL
University of London
King's College
68, Half Moon Lane
London SE24 9JF

Mr. J.D.L. HARRISON
Energy Technology Division
Building 156
AERE, Harwell
Didcot, Oxfordshire

Mr. G. LEACH
International Institute for Environment and
Development
10, Percy Street
London W1P ODR

Dr. B. McNELIS
General Technology Systems Ltd.
Forget House
20 Market Place
Brentford, Middlesex TW8 8EQ

Dr. P.J. MUSGROVE
Reading University
Department of Engineering
Whiteknights
Reading Berks RG6 2AY

Prof. J.K. PAGE
University of Sheffield
Department of Building Science
Western Bank
Sheffield S10 2 TN

Sir George PORTER
Director
The Royal Institution
21, Albemarle street
London W1X 4BS

Mr. F.C. TREBLE
43, Pierrefondes Avenue
Farnborough, Hants GU14 8PA

Mr. D.P. TURNER
Ministry of Overseas Development
Eland House, Stag Place
London SW1E 5DH

Dr. D.J. WHITE
Ministry of Agriculture
Great Westminster House
Horseferry road
London

UPPER-VOLTA

Mr. KABORÉ
Direction de l'H.E.R.
Place de l'Indépendance
Ouagadougou

Mr. P. SOUBEIGA
Voltelec
Binger
Ouagadougou

Mr. G. VERIDIQUE
Directeur de l'Ecole Inter-Etats d'Ingénieurs de
l'Equipement Rural
B.P. 7023
Ouagadougou

U.S.A.

Mr. Turbi ABDU
Arabic Trade Consulting
P.O. Box 11455
Milwaukee, Wis. 53211

Mr. W. CORCORAN
Department of Energy
20 Mass. Ave. (M.S. 2221 C)
Washington D.C. 20545

Mr. P.G. PATIL
Systems Consultants, Inc.
1054, 31st street, NW
Washington D.C. 20007

Mr. W. PORTER
US Department of Energy
Room 3H017
1000 Independence ave. SW
Washington DC 20545

VENEZUELA	Mr. N. GRILLET
	Ministry of Energy and Mines
	Centro Simon Bolivar
	Caracas

Ing. A. PEREZ
Ministry of Energy and Mines
Head of Division of Non-Conventional Energy
Centro Simon Bolivar
Caracas

YOUGOSLAVIA
Mr. N. BOGDANOVIC
Blatine 46
Split

ZAIRE
Citoyen K.N. ISENGINGO
Service Présidentiel d'Etudes
Lokole.4
Kinshasa

Citoyen K. MATANGILA
Service Présidentiel d'Etudes
Lokole 4
Kinshasa

ZAMBIA
Mr. R.G. MITI
General Manager
Kariba North Bank
P.O. Box RW 194
Lusaka

INTERNATIONAL ORGANISATIONS

UNITED NATIONS
New York 10017/USA

Mr. D. LOVEJOY

UNCSTD
(United Nations Conference on Science & Technology for Development)
P.O. Box 274
1015 Vienna/Austria

Mr. J.P. SEKA

FAO
(Food and Agricultural Organisation of the United Nations)
Via delle Terme de Caracalla
00100 Roma/Italy

Mr. JAMES
Mr. H.J. VAN HULST

UNESCO
(United Nations Educational, Scientific and Cultural Organisation)
7, place de Fontenoy
75700 Paris/France

Mr. J.F. McDIVITT

UNIDO
(United Nations Industrial Development Organisation)
Lerchenfelderstr. 1
P.O. Box 707
1070 Vienna/Austria

Mr. P. KOENZ
Mr. G. KOMISSAROV

WMO
(World Meteorological Organization, Industrial Applications and Climatology Branch of the Meteorological Application & Environment Dept.)
41, Av. Giuseppe Motta
B.P. 5
1211 Geneva/Switzerland

Miss S. JOVICIC

REGIONAL ORGANISATIONS

COMMISSARIAT INTER-ETAT POUR LA LUTTE CONTRE LA SECHERESSE (CILSS)
Ouagadougou/Haute-Volta

Mr. L. RANGER

COMMUNAUTE ECONOMIQUE DE L'AFRIQUE DE L'OUEST (CEAO)
B.P. 643
Ouagadougou/Haute-Volta

Mr. M. KANE

INSTITUT DU SAHEL
Bamako/Mali

Mr. O. KANE
Directeur Général

FEDERATION OF ARAB COUNCILS FOR SCIENTIFIC RESEARCH
P.O. Box 13027
Baghdad/Iraq

Prof. M.E. ABBAS

INTER-AMERICAN DEVELOPMENT BANK
808, 17th Street NW
Washington DC 20577/USA

Mr. DA SILVA

LATIN AMERICAN ENERGY ORGANISATION (OLADE)
Casilla 119-A
Quito/Ecuador

Mr. G. RODRIGUEZ ELIZA-RRARAS

EUROPEAN PARLIAMENT

Bâtiment Robert Schuman
Plateau du Kirchberg
Luxembourg

Mr. G. FLÄMIG
Vice-Chairman
Committee on Energy and Research
Bundeshaus
5300 Bonn/Germany

Mr. BERSANI
Vice-Chairman
Committee on Development and Cooperation
Via Lame, 118
40136 Bologna/Italy

ECONOMIC AND SOCIAL COMMITTEE

Rue Ravenstein 2
1000 Bruxelles/Belgium

Prof. P. HATRY
Président de la Section de l'énergie et des questions nucléaires du C.E.S. c/o Fédération Pétrolière Belge
Rue de la Science
1040 Bruxelles/Belgium

COMMISSION OF THE EUROPEAN COMMUNITIES

Dr. G. BRUNNER, Member of the Commission

CABINET G. BRUNNER

Mr. J. KÜHN

DG I — EXTERNAL RELATIONS

Mr. G. BEINHARDT

DG III — INTERNAL MARKET AND INDUSTRIAL AFFAIRS

Mr. A. ANFOSSI

DG VIII — DEVELOPMENT

Mr. FOLEY, Deputy Director-General

Messrs. BINI-SMAGHI, CORNELLI, COX, DE HOE, HECQ, LAIDLER, LEQUEUX, LIVI, REITHINGER, VINCENT

DG XII — RESEARCH, SCIENCE AND EDUCATION	Mr. G. SCHUSTER, Director-General
	Messrs. BATTI, PALZ, STEEMERS, STRUB, VALENTINI, CAPRIOGLIO (Expert)
DG XIII — SCIENTIFIC AND TECHNICAL INFORMATION AND INFORMATION MANAGEMENT	Mr. D. NICOLAY
DG XVII — ENERGY	Messrs. KAUT, MARTIN, RENAUD
JOINT RESEARCH CENTRE	Mr. S. VILLANI, Director-General
	Messrs. ARANOVITCH, BECKER, BEGHI, BONNAURE, GRETZ, FINZI, KREBS, KLERSY, SLESSER

COMMISSION DELEGATIONS TO ACP COUNTRIES

ETHIOPIA	Mr. W. VAN WOUDENBERG
NIGER	Mr. J.P. MARTIN
SOMALIA	Mr. Ch. PELLAS
TCHAD	Mr. J.G. COLLET
ZAIRE	Mr. E. PÖRSCHMANN
GROUPE DES ETATS D'AFRIQUE DES CARAIBES ET DU PACIFIQUE (ACP) Avenue Georges Henri 451 1040 — Bruxelles/Belgique	Mr. L. OUANGRAOUA
CENTRE DE DEVELOPPEMENT INDUSTRIEL (ACP) Avenue Georges Henri 451 1040 — Bruxelles/Belgique	Mr. A. A. ARMANI